Dynamics of Big Internet Industry Groups and Future Trends

Miguel Gómez-Uranga
Jon Mikel Zabala-Iturriagagoitia
Jon Barrutia
Editors

Dynamics of Big Internet Industry Groups and Future Trends

A View from Epigenetic Economics

 Springer

Editors
Miguel Gómez-Uranga
Department of Applied Economics
University of the Basque Country,
   UPV/EHU
Bilbao
Spain

Jon Mikel Zabala-Iturriagagoitia
Deusto Business School
University of Deusto
Donostia-San Sebastian
Spain

Jon Barrutia
Department of Management and Business
   Economics
University of the Basque Country,
   UPV/EHU
Bilbao
Spain

ISBN 978-3-319-80975-5   ISBN 978-3-319-31147-0   (eBook)
DOI 10.1007/978-3-319-31147-0

Printed on acid-free paper

This Springer imprint is published by Springer Nature
The registered company is Springer International Publishing AG Switzerland

In Memoriam,
Miguel Gómez-Uranga
26/09/1948–17/12/2015

Preface

The story of this book dates back to 2012. At that time, Jon Mikel Zabala-Iturriagagoitia was working at the Centre for Innovation, Research and Competence in the Learning Economy (CIRCLE), Lund University (Sweden), studying knowledge-based entrepreneurship in the country's software industry. At the same time, Miguel Gómez-Uranga had started to read about symbiotic processes and new findings from molecular biology. In turn, Jon Barrutia was focused on the physical properties of processes at the nanoscale. In one of the visits Jon Mikel made to Bilbao (Spain), the three editors of the book exchanged ideas, discussed their findings, and began talking about a new concept: epigenetics.

In January 2013, Jon Mikel moved to a new position at the University of Deusto in San Sebastian (Spain). As a result, the three editors undertook joint research on epigenetics and how the concept could be brought into the study of innovation at the organizational level. In particular, we were interested in exploring the Internet ecosystem, due to the dramatic innovation processes occurring in it. After a year of research, in 2014 we introduced the concept of "Epigenetic Economic Dynamics" in a paper published in Technovation.

A few months later, we got an e-mail from Anthony Doyle, engineering editor at the UK office of Springer, who invited us to write a book delving into the concept of Epigenetic Economic Dynamics (EED). We were stunned by the offer, which derived in careful thought on whether to accept the challenge or not. We were a small team, and even though a book would allow us to expand and improve our ideas, we were very much aware that writing a book differs greatly from writing a paper. After long discussion among the three of us, we decided to take a step forward and provided Springer with an official proposal for this book in July 2014. The contract with Springer was signed in October 2014 and the 'book project' was underway.

In order to complete the book, we felt the need to include a wide array of competences and perspectives on the evolution of the Internet ecosystem. Its singularity, its light-speed evolution, and the business diversity found in it demanded an interdisciplinary approach. It was necessary to take the scope of the book beyond EED focus and gain inputs from other specialist colleagues in the areas we intended to study. In December 2014 we organized a workshop at the University of the Basque Country in Bilbao, to which several researchers from different Spanish universities were invited to present their views on the Internet ecosystem and their potential contribution to the book. We consider this to have been the kick-off meeting for this book.

This interdisciplinary work obliged us to be knowledgeable about systemic and managerial approaches to innovation. The contribution of computer and telecommunications engineers was required to gain an understanding of the technological phenomena. Knowledge of strategic management was key in explaining the rational behavior of the large business groups we were interested in examining. At the same time, we were interested in gaining the commitment of developers. Developers are becoming the cornerstone for the rapid development of the Internet and the dramatic growth of the large industry groups dominating it. We needed practice-oriented developers, who could provide new insight into how the Internet ecosystem is seen through their entrepreneurial eyes. Combining all these academic and practice-oriented cultures, as well as individuals with different levels of dedication, has not been an easy task. However, our aim to provide a quality product with a high degree of novelty has overcome all the difficulties that we have encountered during the preparation of this book.

As in any human endeavor, we have been through situations, as could be expected. There have, however, been many more circumstances that we did not foresee and that we cannot draw conclusions from. The first were the very serious illnesses suffered by Miguel Gómez-Uranga and Goio Etxebarria, which put the 'book project' on hold for a while. The excellent care received from the oncologists and hematologists who treated them kept their conditions from becoming an impediment; not to mention the special role played by their families, friends, and colleagues. Adversity can sometimes help us to give the best of ourselves.

Timing has also been an issue, and sometimes we felt that we were confusing two terms that, despite looking similar, have very different implications: importance and emergency. We have experienced, suffered, and endured processes that even made some people consider the possibility of some of the authors appropriating intellectual property they did not own. Basque people are said to be characterized by their clarity, transparency, and the value they place on a given word. Maybe, as in epigenetics, our Basque DNA has been altered by the context. This often meant that when we were discussing the progress of the book, and exchanging our views on how the process should be led, we ended our discussions with an "NN" (we to ours—or "nosotros a lo nuestro" in Spanish). For all of the above, we must express our gratitude to Garrett Ziólek, editorial assistant at Springer UK, for his understanding and flexibility with regard to all the administrative work needed for the edition of this book.

We are finally able to deliver this 'eclectic' book. During this year, we have often seen ourselves not as great chefs of the worldwide recognized Basque haute cuisine, but more like those small restaurant cooks now fashionable on TV shows which have to do something with what they have readily available, and for which time is short. In our case, the final opinion is the reader's.

We do not want to forget the role and contribution of those who have helped us improve the quality of the manuscripts, be it translating them into English or reviewing the editing of the text. In this regard we are particularly indebted to Patricia O'Connor and Aimar Basañez Zulueta, without whom this endeavor would have not been possible. Funding is also a key factor when carrying out an activity of this type. We are thus grateful to the Basque Government for the funding provided through the Etortek 2014 program to the Nanogune' 14 project.

Bilbao, Spain   Miguel Gómez-Uranga
Donostia-San Sebastian, Spain   Jon Mikel Zabala-Iturriagagoitia
Bilbao, Spain   Jon Barrutia
October 2015

PS: On 17 December 2015 Miguel Gómez-Uranga passed away after a long disease. This book constitutes a memory, a tribute, and a recognition for his hard work on the development of a comprehensive analytical contribution based on the concept of "Epigenetic Economic Dynamics" during these past years. Rest in peace.

Contents

Introduction 1
Miguel Gómez-Uranga, Jon Mikel Zabala-Iturriagagoitia and Jon Barrutia

**Introducing an Epigenetic Approach for the Study
of Internet Industry Groups** 13
Miguel Gómez-Uranga, Jon Mikel Zabala-Iturriagagoitia and Jon Barrutia

Epigenetic Economics Dynamics in the Internet Ecosystem 53
Jon Mikel Zabala-Iturriagagoitia, Miguel Gómez-Uranga,
Jon Barrutia and Goio Etxebarria

**GAFAnomy (Google, Amazon, Facebook and Apple):
The Big Four and the b-Ecosystem** 127
Juan Carlos Miguel and Miguel Ángel Casado

The Digital Ecosystem: An "Inherit" Disruption for Developers? 149
Jorge Vega, Jon Mikel Zabala-Iturriagagoitia
and José Antonio Camúñez Ruiz

Future Paths of Evolution in the Digital Ecosystem 179
Jorge Emiliano Pérez Martínez and Silvia Serrano Calle

4G Technology: The Role of Telecom Carriers 201
Andrés Araujo and Itziar Urizar

**Scope and Limitations of the Epigenetic Analogy:
An Application to the Digital World** 243
Jon Barrutia, Mikel Gómez-Uranga and Jon Mikel Zabala-Iturriagagoitia

Editors and Contributors

About the Editors

Miguel Gómez-Uranga was full Professor in the Department of Applied Economics I, University of the Basque Country in Bilbao (Spain), where he lectured and conducted research for more than thirty years. His research interests included economics, technical change, and regional economics. He was also involved in the development of institutional and evolutionary economics and was visiting professor at the University of Lyon, the University of Nevada, and the University of Melbourne, among others. During his long career, he served as coordinator and researcher for the Basque Country in several research projects financed by the European Commission. Some of the most emblematic ones include the "VALUE Program," "Regional Innovation Systems: designing for the future" and "Universities, technology transfer and spin-off activities: Academic entrepreneurship in different types of European Regions." His seminal work on Regional Innovation Systems, together with Professors Phil Cooke and Goio Etxebarria, is regarded as one of the key scholarly contributions in the innovation studies community. His extensive publication list includes several books, books chapters, as well as papers in journals such as Research Policy, Environment and Planning A, Journal of Economic Issues, European Planning Studies, Technovation or Scientometrics. RIP.
Email: miguel.gomez@ehu.es

Jon Mikel Zabala-Iturriagagoitia is lecturer at the University of Deusto in San Sebastian (Spain). He was previously Assistant Professor at the Centre for Innovation, Research and Competence in the Learning Economy (CIRCLE), Lund University (Sweden). His research and teaching interests are related to the fields of innovation policy and innovation management. As a researcher, he has contributed to the development of methodological approaches for the assessment of innovation potential, innovation policy instruments such as public procurement for innovation and pre-commercial procurement, and innovation management tools (in particular, creativity and technology watch). As a lecturer, he has been engaged in courses at the Ph.D., Master's degree, and undergraduate levels at several European and Latin

American universities. His work has been published in journals such as European Planning Studies, Regional Studies, Scientometrics, Research Policy, Science and Public Policy, R&D Management, Technovation, Review of Policy Research or Entrepreneurship and Regional Development, among others.
Email: jmzabala@deusto.es

Jon Barrutia is Professor of Management and Business Economics at the University of the Basque Country in Bilbao. He has extensive academic and professional experience in various high-level academic and administrative positions. Between 1990 and 1995 he was Vice Dean of the Faculty of Economics and Business Administration at the University of the Basque Country. Between 1996 and 1999 he was Vice Rector of the University of the Basque Country. In 1999 he was appointed as a Director of Universities in the Basque Government, a position he held until 2001. He then held the office of Vice Councillor of University and Research of the Basque Government between 2001 and 2006. Since 1987 he has been a member of the European Academy of Management and Business Economics, and since 2009 he is Head of the Department of Management and Business Economics at the University of the Basque Country. As a lecturer, he has taught courses related to organizational theory and public management at the undergraduate, executive education, MBA, and Ph.D. levels.
Email: jon.barrutia@ehu.es

Contributors

Andrés Araujo is Professor of Financial Economics and Accounting at the Faculty of Economics and Business Administration, University of the Basque Country (UPV/EHU). His main areas of research include management of innovation and entrepreneurship. He lectures at the undergraduate level in areas such as Strategic Management, Innovation Management and Public Enterprise Management. At the Master's degree level he also lectures on various executive MBAs, the Master's degree in Marketing, the Master's degree in Internationalization and Innovation Research, and the MBA on Entrepreneurship. Between 2009 and 2012 he served as Deputy Director for the Economy, Budget and Economic Control of the Basque Government. He is now a member of the Consolidated Research Group on Built Heritage. He was formerly a member of the Global Entrepreneurship Monitor (GEM) of the Basque Country, Head of the Institute of Applied Economics on Business (University of the Basque Country), and Chief Commissioner of the School of Business Studies in Vitoria. In 2013 he was visiting researcher at the University of Manchester (UK). His publication list includes several papers in journals such as the Spanish Journal of Finance and Accounting, the Chinese Business Review, Información Comercial Española, Revista de Economía Mundial, Maritime Policy and Management, Human Factors and Ergonomics in Manufacturing & Service Industries, or Applied Economics Letters among others.
Email: andres.araujo@ehu.es

José Antonio Camúñez Ruiz is Associate Professor at the Department of Applied Economics I, University of Sevilla (Spain). His area of specialization is quantitative methods for economics and business studies. In this regard, to date, he has coauthored five books, more than 30 book chapters, and more than 30 journal papers. Currently, he is in charge of the development of a statistical technical report on tax inspection processes through indirect estimation. Besides, he is also engaged in other research projects and contracts focused on the evaluation of innovative programs for professional development at universities, the development of a system of price indices for goat milk, and the identification of potential indicators that favor reasoned proposals for price updating. In addition, he is involved in lecturing on a variety of subjects such as mathematics, statistics, measures of inequality, and research methodology at the faculties of economics and business of the University of Sevilla and at the school of tourism and finance of the same university.
Email: camunez@us.es

Miguel Angel Casado is vice-dean of the Faculty of Social Sciences and Communication of the University of the Basque Country, where he has been researcher since 2003. He is lecturer at the Department of Audiovisual Communication and advertising where he teaches the course "Internet and the cultural industries" and also teaches the course "tools for qualitative research" at Social Communication Research Master of the University of the Basque Country. His research interests include cultural industry economics, public media systems, and children and new media. He has been member of the European research network funded by the European Commission EU KIDS ONLINE and now is involved on different research projects funded by the Spanish Ministry of Innovation.
Email: miguelangel.casado@ehu.eus

Goio Etxebarria is full Professor at the Department of Applied Economics I, University of the Basque Country in Bilbao (Spain). His research has been mainly focused on regional economic development and bibliometrics. During the 1990s he was visiting researcher at the University of Leeds in the United Kingdom. After returning to the University of the Basque Country, he participated in several EU-wide research projects, such as "Declining Industry and Regional Regeneration in Wales and the Basque Country," "Regional Technological Centres in the Basque Country," "Regional Innovation Systems: Designing for the future" or "Integrating Technological and Social Aspects of Foresight in Europe." His seminal work on Regional Innovation Systems, together with Professors Phil Cooke and Miguel Gómez-Uranga, is regarded as one of the key scholarly contributions in the innovation studies community. His extensive publication list includes several books, books chapters as well as papers in journals such as Research Policy, Environment and Planning A, Journal of Economic Issues, European Planning Studies or Scientometrics.
Email: goio.etxebarria@ehu.es

Juan Carlos Miguel is full Professor in the Department of Audiovisual Communication at the University of the Basque Country in Bilbao (Spain). He has lectured and studied Economics of the Media and Entertainment for more than twenty years.

His research focuses on the strategies of global media and entertainment groups, as well as prices and free content on the Internet. He has been the Director of the Ph.D. Program on the Internet and Cultural industries. He has been visiting professor at several universities such as Glasgow University, Grenoble University, Paris III University, and Quiles University (Argentina). He is Associate member of the Center for research on social innovations (CRISES—Centre de recherche sur les innovations sociales) at Quebec University in Montreal.
Email: jc.miguel@ehu.eus

Jorge Emiliano Pérez Martínez is full Professor in the Department of Signals, Systems and Radiocommunications at the Telecommunications Technical School (ETSIT)—Polytechnic University of Madrid (UPM). He holds a Ph.D. in Telecommunications Engineering from the ETSIT-UPM and a Master's degree in Social and Political Science from the Complutense University of Madrid. Dr. Pérez leads the Research Group in Communications and Information Technologies (GTIC-SSR-UPM). From 1990 to 1999 he was President of the Spanish Professional Association of Telecommunications Engineers and Dean of the Spanish Telecommunications Engineers. From September 2003 to June 2004 he was General Director for the development of the Information Society at the Ministry of Science and Technology in Spain. He has been the Chair on Information Society Development of Red.es where he coordinated the Spanish Experts Group on Analysis and Prospective on Telecommunications (GAPTEL). Since the Spanish Internet Governance Forum (IGF Spain) was created, Dr. Pérez has been its coordinator. In 2015 he was appointed Director of Digital Economy at the public institution Red.es. His research interests are focused on new advanced telecommunication services and telecommunication policy and regulation. He is the author of many books and a large number of academic papers that have been published in Spanish and international journals. For many years Dr. Pérez has successfully coordinated major national and international research projects financed by private and public institutions.
Email: Jorge.perez.martinez@upm.es

Silvia Serrano Calle is Associate Professor and researcher at the Polytechnic University of Madrid (UPM). Ph.D. in Economics from the Spanish Open University (UNED), she holds a degree in Telecommunications Engineering from the Telecommunications Technical School (ETSIT)-UPM and an Economics degree from the UNED. She has lectured and conducted research at the UNED, as Associate Professor in the Department of Economics at the Pontifical University of Comillas in Madrid, and currently at ETSIT-UPM. She is a member of the Research Group on International Political Economy and Energy (UNED) and of the Research Group in Communications and Information Technologies (GTIC-SSR-UPM). Dr. Serrano is Technical Director of the Spanish Internet Governance Forum. She is a member of the Royal Spanish Physical Society (RSEF) and in 2015 was elected a member of the RSEF's Executive Committee. Dr. Serrano has combined her academic dedication with a professional career in the private sector, where she has held technical and management positions in multinational companies with responsibilities over technical departments, investment, marketing, and sales. Dr. Serrano has

written a book on regulation and regulatory risk in the energy sector published by the Escuela de Organización Industrial (EOI—Spanish Ministry of Industry, Energy and Tourism) and research papers and book chapters on regulation, public policy, telecommunications, internet governance, energy economics, and risk assessment.
Email: Silvia.serrano@upm.es

Itziar Urizar joined the University of the Basque Country (UPV/EHU) in 1998. She is Associate Professor in the Department of Applied Economics I at the Faculty of Economics and Business Administration in Bilbao (Spain). She is currently teaching Economic Environment Analysis to undergraduate students at the Faculty of Social Sciences and Communication of the UPV/EHU. Additionally, she is participating in the Master's degree on "Management of Human Resources and Employment" at the School of Industrial Relations of the University of the Basque Country. She has formerly taught courses such as Introduction to Economics, Economic Policy or Mathematics. She has been engaged in the Ph.D. program on "Globalization, Development and Cooperation" at the UPV/EHU. Her research interests lie in the field of national and regional innovation systems, and science, technology, and innovation policies. She has authored several articles in journals, the most recent of which has been recently published in the Center for Basque Studies at the University of Nevada, Reno.
Email: itziar.urizar@ehu.es

Jorge Vega is "front-end" developer at Merkatu interactiva, a company that provides consulting services in e-business projects and digital management. His passion and main occupation lie in software development. Furthermore, he is interested in investigating the inclusion of new technologies in organizations, as well as their use and development in institutional, social, and business contexts. He has more than ten years' experience in the development of web-based projects. These include projects related to e-commerce, multimedia, and corporate projects, which have been developed in and deployed for different organizations, industries, and environments. Between 2005 and 2007 he was a developer at Kubik Web Studio, a company created as a result of his cooperation with other entrepreneurs. He had previously worked as a programmer at Dinamatik Web Studio. He was a professional athlete for more than 13 years, which taught him the value of perseverance and continuous effort and work in order to achieve long-term goals.
Email: jvega300@gmail.com

List of Figures

Introduction

Figure 1   New research horizons that may open as a result
of the EED approach. *Source* Own elaboration.   5

**Introducing an Epigenetic Approach for the Study of Internet
Industry Groups**

Figure 1   Time and substantial change in the environment.
Source Own elaboration.   25

Figure 2   The epigenetic economic dynamics (EED) approach.
Source Own elaboration.   40

Epigenetic Economics Dynamics in the Internet Ecosystem

Figure 1   The map of the Internet. *Source* http://orig01.deviantart.net/91af/
f/2014/070/a/5/map_of_the_internet_2_0__by_
jaysimons-d781bst.jpg. Accessed 19 December 2015.   62

Figure 2   Number of employees (FTE). *Source* Own elaboration   74

Figure 3   Sales (revenues)—Million US$. *Source* Own elaboration   74

Figure 4   Sales (revenue) per employee—Million US$. *Source*
Own elaboration.   75

Figure 5   Gross profit—Million US$. *Source* Own elaboration   76

Figure 6   Gross profit per employee—Million US$. *Source*
Own elaboration.   76

Figure 7   R&D investments—Million US$. *Source* Own elaboration   77

Figure 8   Share of R&D investments on sales. *Source* Own elaboration . . .   78

Figure 9   Number of USPTO patents granted. *Source* Own elaboration . . .   78

Figure 10   USPTO patents granted per billion US$ invested
on R&D. *Source* Own elaboration.   79

Figure 11   USPTO patents granted per 1000 employees.
Source Own elaboration.   79

Figure 12   Market capitalization (billion US$). *Source* Own elaboration . . .   80

Figure 13   **a** Google's ecosystem (2008). *Source* Iyer and Davenport
             (2008). **b** The core of Google's Global Innovation
             Ecosystem (2014). *Source* Fransman (2014: 24).   82
Figure 14   **a** Apple's ecosystem (2011). *Source* TIME,
             September 12, 2011. Available: http://obamapacman.com/
             2011/09/time-magazine-apple-ecosystem-infographic/.
             b Apple's innovation ecosystem (2014). *Source* Nielson (2014). . . .   85
Figure 15   Facebook's ecosystem (2011). *Source* Trewe (2011)   86
Figure 16   The evolution of Amazon's business model.
             Source Isckia and Lescop (2009: 45).   88
Figure 17   **a** Microsoft's business ecosystem. *Source* Skelly (2014).
             b The core of Microsoft's Global Innovation Ecosystem.
             Source Fransman (2014: 27)   90
Figure 18   Twitter's business ecosystem. *Source* Bmimatters (2012).   91
Figure 19   Number of USPTO patents granted between 1984
             and 2014. *Source* Own elaboration based on USPTO.   105
Figure 20   **a** Number of yearly USPTO patents granted to some
             of the large business groups operating on the Internet
             (2007–2014). *Source* Own elaboration based on USPTO.
             b Number of yearly USPTO patents granted to some
             of the large business groups operating on the Internet
             (2007–2014). *Source* Own elaboration based on USPTO.   106
Figure 21   Technological diversification path followed by Twitter
             (2013–2014). *Source* Own elaboration based on USPTO.   108
Figure 22   Technological diversification path followed by eBay
             (2000–2014). *Source* Own elaboration based on USPTO.   108
Figure 23   Technological diversification path followed by Amazon
             (1998–2014). *Source* Own elaboration based on USPTO.   109
Figure 24   Number of M&As completed (1987–2015)
             (We could not find any evidence of the M&As
             completed by Samsung.). *Source* Own elaboration.   111
Figure 25   Value of the completed M&As (1987–2015)
             (Million US$). *Source* Own elaboration   112
Figure 26   **a** Number of M&As completed (1987–2015) by some
             of the large business groups operating on the Internet.
             Source Own elaboration. **b** Number of M&As completed
             (1987–2015) by some of the large business groups operating
             on the Internet. *Source* Own elaboration   113
Figure 27   **a** Value of the completed M&As (1987–2015) by some
             of the large business groups operating on the Internet
             (Million US$). *Source* Own elaboration. **b** Value
             of the completed M&As (1987–2015) by some of the large
             business groups operating on the Internet (Million US$).
             Source Own elaboration.   115

Future Paths of Evolution in the Digital Ecosystem

Figure 1   The digital ecosystem's value chain. *Source*
            Own elaboration based on Kearney (2010)   188

4G Technology: The Role of Telecom Carriers

Figure 1   Share (%) of Internet mobile device traffic. *Source*
            Own elaboration based on stat counter global stat data.
            Note (*) data up to April 2015   202
Figure 2   Forecast of mobile data traffic. *Source* Cisco visual
            networking index. Gerhardt et al. (2013).   202
Figure 3   Revenues of mobile telephony in Spain. *Source* Own
            elaboration based on data from the Spanish national
            telecommunications market commission.   219
Figure 4   Mobile lines in Spain. *Source* Authors,
            from CNMC (2014a, b)   219
Figure 5   Italia Telecom. *Source* Prepared by authors using data
            from Italia telecom financial reports   223
Figure 6   Revenues of SMS in Spain. *Source* Own elaboration
            based on CNMC (2014a, b).   224

**Scope and Limitations of the Epigenetic Analogy: An Application
to the Digital World**

Figure 1   Reframing the EED approach. *Source* Own elaboration   253

List of Tables

Introducing an Epigenetic Approach for the Study of Internet Industry Groups

Table 1   The timing of adaptation in complex systems   24
Table 2   Some features necessary to understanding epigenetic dynamics. . . .   31
Table 3   Summary of views on routines seen from the EED perspective. . . .   36
Table 4   An epigenetic approach to understanding the economic
          and social impact of ecosystem dynamics   45

Epigenetic Economics Dynamics in the Internet Ecosystem

Table 1   A conceptualization of the Internet economy   65
Table 2   The origins of the GAFAs   72
Table 3   An epigenetic understanding of the economic impact
          of big internet business ecosystems' dynamics   93
Table 4   The companies' activities in different fields (market shares
          and ranking by activities or products) between 2012 and 2013. . . .   95
Table 5   Economic dynamics in some of the leading Internet
          business groups in 2014   96
Table 6   Cyber challenges and dangers faced by the GAFAs
          and other SMEs   102
Table 7   Possible measures to face the cyber risks.   103
Table 8   Methodology followed to assess the technological diversification
          of the GAFAs. *Source* Own elaboration.   107
Table 9   Classic and constructive assessment of technologies.   119
Table 10 Standard and constructivist view on technology   120

GAFAnomy (Google, Amazon, Facebook and Apple): The Big Four and the b-Ecosystem

Table 1   GAFA. Some financial information. In dollars, units
          (employees) and % (RD/sales).   132
Table 2   GAFA's investments in acquisitions from 2011 to 2014
          ($Millions)   133

Table 3   GAFA. Competitors according to item 1A of the 10-K
           (risk factors)   134
Table 4   Competition in GAFA, by business activities   136
Table 5   GAFA. Examples of platforms.   140

The Digital Ecosystem: An "Inherit" Disruption for Developers?
Table 1   History and influence of programming languages   156
Table 2   Relationship between the quantum properties
           and developer dynamics   175

4G Technology: The Role of Telecom Carriers
Table 1   Mobile telephone sector's contribution to GDP (2014)
           and growth rates (2009–2014)   203
Table 2   Market share of mobile broadband in Spain (2012).   211
Table 3   Leaders in 4G patent-related issues   216
Table 4   Market share of mobile carriers in the USA   220
Table 5   Payout ratio and market to book value in selected carriers
           and OTT companies (2014)   223

Abstract

The aim of this book is to contribute to the literature on innovation studies with new findings from biogenetics that are becoming increasingly important. In particular, we discuss the new analytical frameworks that may open up as a result of the inclusion of epigenetics in evolutionary economic thinking. The book introduces a new approach to the study of business dynamics, namely Epigenetic Economic Dynamics. The concept of Epigenetic Economic Dynamics (EED) is defined as the study of the (epigenetic) dynamics generated as organizations adapt to major changes in their respective environments.

The EED approach is illustrated by studying the evolution of big Internet industry groups such as Apple, Google, Microsoft, Facebook, Amazon, and Samsung. The EED approach enables us to understand how the dynamics of these groups address changes in their environments. It is also useful when analyzing the results or consequences of these dynamics. Abnormalities, malfunctions or obstacles to innovation, and/or blockage to developing competition at certain levels (i.e., intellectual property rights, abuse of monopoly power, etc.) may arise as a result of the influence of epigenetic dynamics.

The book provides a better understanding of the rapid change dynamics that are taking place in big Internet industry groups. It enables us to capture the dynamics and changes through a coherent and consistent model, particularly in relatively short periods of time (i.e., highly dynamic environments), which are those occurring within and between the main business groups operating on the Internet, as well as within and between telecom operators (i.e., telecom carriers).

The book targets a wide audience, ranging from scholars and practitioners interested in innovation studies, the implementation and management of technology and innovation, management studies, business administration and innovation policies, to postgraduate students of Economics or Engineering.

Introduction

Miguel Gómez-Uranga, Jon Mikel Zabala-Iturriagagoitia and Jon Barrutia

1 Introduction

The aim of this book is to contribute to the literature on innovation studies with new findings from biogenetics that are becoming increasingly important. In particular, we will discuss the new analytic frameworks that may open as a result of the incorporation of epigenetics in evolutionary economic thinking. The book introduces a new approach to the study of business dynamics, namely Epigenetic Economic Dynamics. The concept of Epigenetic Economic Dynamics (EED) is defined as the study of the (epigenetic) dynamics generated as a result of the adaptation of organizations to major changes in their respective environments (Gómez-Uranga et al. 2014).

We are witnessing a very rapid development of mobile telephone-related sectors and technologies on the Internet in recent years. The exponential growth of users puts such pressure on the demand that it leads to major changes in the dynamics of the environment, prompting a huge business and thousands of new user applications. A few groups dominate this fast paced panorama (i.e., Apple, Google, Microsoft, Facebook, Amazon, Samsung, Alibaba to mention a few), with their behavior determining the direction and intensity of global dynamics.

M. Gómez-Uranga
Department of Applied Economics I, University of the Basque Country,
UPV/EHU, Bilbao, Spain
e-mail: miguel.gomez@ehu.es

J. Barrutia
Department of Management and Business Economics, University of the Basque Country,
UPV/EHU, Bilbao, Spain
e-mail: jon.barrutia@ehu.es

J.M. Zabala-Iturriagagoitia (✉)
Deusto Business School, University of Deusto, Donostia-San Sebastian, Spain
e-mail: jmzabala@deusto.es

© Springer International Publishing Switzerland 2016
M. Gómez-Uranga et al. (eds.), *Dynamics of Big Internet Industry Groups and Future Trends*, DOI 10.1007/978-3-319-31147-0_1

The book also explores a related technological domain, the one involving 4G technology providers, namely large telecom operators (i.e., carriers). This book centers on finding how we can understand these dynamics and thus have a clearer view of their evolution. With that, we shed light on the dynamics of business groups, which we approach as "business ecosystems" (Moore 2005; Razavi et al. 2010).

4G technology is revolutionizing the evolution of the Internet, which in turn has very important consequences on various aspects. First of all, the need to allow the movement of large masses of information implies on the one hand that all platforms must meet and conform to these requirements, while on the other hand, large telecom operators (i.e., carriers) need to offer network infrastructures able to support exponential increases in the amount of data under circulation. Second, and as a consequence, there is an increasing need to develop new standards that allow the previous context to develop, which in turn generates new patenting dynamics by the main business groups, both in the telecommunications sector and in the business groups that operate on the Internet. It is worth noting the very dissimilar dynamics that are being observed in the different geographical regions (i.e., groups of countries) in relation to the time of implementation of 4G technology. For example, if Japan or the United States is compared with Europe, in the former, there is a much larger deployment of these technologies. In this regard, "wobbling" dynamics can be observed in the deployment of adequate networks by telecommunication operators and Chinese Internet groups.

Epigenetics, in the field of molecular biology, has shown the phenotype can vary as a result of long-term and short-term adaptation (Francis 2011; Carey 2012). The EED approach, created to explain dynamics observed in highly dynamic environments, also considers that some decisions are or need to be formulated in other stable environments. As the book discusses, the decision-making processes that aim at guaranteeing adaptation and survival in both short-term (radical adaptation) and long-term (gradual adaptation) show different characteristics.

The EED approach enables us to understand how the dynamics of the large Internet industry groups mentioned address and respond to changes in their environments. Second, it is also useful when analyzing the results of these dynamics. Abnormalities, malfunctions or obstacles to innovation, and/or blockage to developing competition at certain levels (i.e., intellectual property rights, abuse of monopoly power, etc.) may arise as a result of the influence of epigenetic dynamics. Acquisition of patent portfolios and patent lawsuits for infringements and violations is one of these malfunctions or consequences. These are particularly quite common, for example in the field of mobile telephony, which clearly shows the fierce competition between business groups (Graham and Vishnubhakat 2013). Essential patent licensing is particularly at the core of legal disputes between the business groups (Allekote and Blumenröder 2010).

The book allows increasing the ability to capture through a coherent and consistent model the dynamics and changes, particularly in relatively short periods of time (i.e., highly dynamic environments), which are those occurring within and between the main business groups operating on the Internet, as well as within and between telecom operators (i.e., telecom carriers). It also provides a better understanding of the rapid change dynamics that are taking place in big Internet industry groups.

The first contribution of the EED approach stems from the fact that the dynamics that are shaping actions through decision would be partially unpredictable, based on a rational and logical path, and might be even mathematically "apprehended" (i.e., through probabilistic models). These characteristics related to unpredictability are not, however, exclusive to the concept of EED. Despite the origin of the methodological framework of the EED relies on a biological analogy to evolutionary economics through the concept of epigenetics, where genes, initially, not only determine the evolution of life (Jablonka and Lamb 2005), similar arguments are also to be found in other disciplines: for example, in the theories of irreversibility (i.e., in environmental sciences), in models of the chaos theory or in economic models with high uncertainty.

Second, the decision-making processes that (may) allow the adaptation to the environment are reflected in epigenetic dynamics, which could not be predicted in response to the original behavior of organizations, or the existing path (i.e., path dependence) in the territory these are based. Therefore, the EED approach may provide a useful concept, which could also be easily applicable in other contexts (sectoral, regional, organizational), as long as it would be possible to identify their epigenetic dynamics. The more dynamic the environment is, (organizational and territorial) dynamics would also be shown as being more breakthrough and falling outside the normal process of gradual evolution that evolutionary economics would predict. This would mean that organizations should in some cases face risks that may result from entering new paths that would seem unrecognizable in relation to their past behavior (i.e., radical adaptation). For example, adopting absolutely novel technologies, finding partners from unrelated geographical areas and sectors, or making acquisitions of assets of all types (i.e., buying companies, buying patent portfolios, attracting talent) as mechanisms to ensure the adaptation to convulse environments.

Third, the researcher who may be interested in using the analytical framework of the EED would need some "sensors" or "seismographs" that allow capturing the emergence of new highly dynamic environments, organizations or even areas of activity, where very sudden and abrupt movements are taking place (e.g., equivalent to having a device to measure the intensity of an earthquake).

Finally, compared to other (non-epigenetic) analog models, the EED approach integrates and allows observing, analyzing, and assessing the potential economic, social, institutional, and even moral consequences of the significant changes arising from the adoption of these epigenetic dynamics. Certain methodological proposals can be derived from the EED approach, which would make it possible to use it in other industries:

1. Adaptation to rapid change in business environments may give rise to epigenetic dynamics. Some of the factors that change in the environment may be related to some environmental factors such as disruptive innovations or changes in techno-scientific paradigms.

2. Epigenetic dynamics in business groups cannot be recognized or foreseen in the initial stages of their evolution.
3. Considering these groups as ecosystems, and evolution from the EED perspective are compatible.
4. A large part of epigenetic dynamics comes from the groups' external relations, such as purchasing external assets, which occurs in the case of patents.
5. Some epigenetic dynamics generate consequences.
6. The consequences call for regulatory intervention, antitrust commissions, better agencies (such as patent offices), public intervention, etc.
7. The consequences can sometimes affect the epigenetic dynamics themselves. For example, excessive patent distortion can lead to changes in patent laws and prevent certain practices.

Although the book is centered on the study of epigenetic dynamics and their consequences, we are aware that the systemic effects of these dynamics are much broader than patents and should also include other areas such as law, regulations, institutions, and even the moral sphere. These additional systemic implications could be studied in future research conducted on this business environment as well as in other areas distinctive for their drastic and rapid changes.

This book may also make it easier to forecast future dynamics on the Internet by proving that a sizeable number of big business groups are veering from their initial paths to take unheard of new directions due to pressure from the competition. This obliges each of these groups to move into very different fields of innovation, using their ability to capture resources and external competences (from other companies).

The book aims to complement existing approaches in the field of evolutionary economics, which are mainly based on the principle of gradual adaptation and natural selection, by introducing the concept of EED. We would like to highlight its theoretical and conceptual nature which aims to contribute to recent work based on evolutionary economics, mainly from the perspective of generalized Darwinism (Aldrich et al. 2008), resilience (Boschma 2015), and related variety (Frenken et al. 2007). The view of the generalized Darwinism relies on the principles of retention, selection, and variety, as vehicles to achieve adaptation (Levinthal and Marino 2013). Therefore, under this perspective adaptation takes place in long-term horizons, through slow and moderate adaptation processes. We believe the book can complement the work conducted by scholars on evolutionary economics (Witt 2014) by applying the analytical framework of EED we put forth.

From our point of view, the field of evolutionary economics could be further complemented with findings emanating from other disciplines such as systems theory, complexity theory, biology, and ecology. Our book is therefore a first step in this direction. It also opens new research horizons and opportunities for scholarship in adjacent areas which will have to be addressed in future research. However, we deem necessary to develop an integral and holistic theory that allows to better link the different streams of research that can help supplement and reinforce the principles of evolutionary economics. Figure 1 illustrates how the book may complement to, be compatible with and show an ability to adapt to particular domains

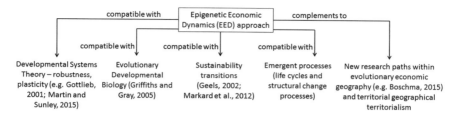

Fig. 1 New research horizons that may open as a result of the EED approach. *Source* Own elaboration

that are increasingly being developed within the field of evolutionary economics. In our view, new theoretical developments such as the EED approach complement, are compatible with and show an ability to adapt the field of evolutionary economics into the highly dynamic environments that are dominating the current economic context.

As shown in Fig. 1, we believe that the contribution outlined by the EED concept may be of interest not only for scholars in the field of evolutionary economics, but also in other related areas such as economic geography, the literature on technology and sustainability transitions, or the research and practice on innovation management (Tidd 2001; Cetindamar et al. 2009) and strategic management (Suárez et al. 2015) through the concepts of dynamic capabilities (Teece et al. 1997; Teece 2012; Eisenhardt and Martin 2000) and organizational routines (Feldman and Pentland 2003). We find certain alignment with some of the developments that are currently taking place and some of the analogies being introduced in the field of environmental sciences in general, and in economic geography in particular. The methodological tools which are normally used in spatial economics and evolutionary economic geography (Boschma and Martin 2007; Martin and Sunley 2007) could be reinforced by the EED approach. It could be useful to explain situations in which the different spaces (national, regional, and local) have to face drastic changes in their environments and in which decision makers would have to implement both short and long term strategies. However, we would also like to emphasize other key concepts from different scientific fields such as emergency (Holland 1998; Corning 2002), which are of relevance to us since emergent change processes (i.e., epigenetic dynamics) are often produced as a response to changes in the environment. In this sense, we find some overlap between Evolutionary Developmental Biology, which is gaining importance, particularly in geographical and spatial studies, and our methodological approach.

However, from our point of view the EED approach finds greatest closeness with developmental systems theory. This approach is mainly the result of the contributions made by environmentalist philosophers such as Gottlieb (2001), Griffiths and Gray (2005) and others such as Oyama et al. (2001) and Robert et al. (2001). Nevertheless, distinguished authors in innovation studies and economic geography, such as Martin and Sunley (2015), introduce and base their contributions on developmental systems theory, when discussing how the expression of the genes should be contextualized. This, in part, reflects the underlying argument of our epigenetic logic. Similarly, the approach on sustainability and technology transitions (Geels 2002, 2014; Markard et al. 2012) can also be complementary to the EED, above all when discussing the characteristics of the environment and its evolution. Thus, we intend that the contribution made by this book can among others, assist, help, and complement the development of the scientific fields mentioned in Fig. 1.

We acknowledge that the methodological approach introduced in this book is quite functional and that is further operationalization remains for further work. Consequently, the three-stage approach the EED is based upon should be understood as a preliminary approximation to the understanding of the underlying dynamics that characterize fast-evolving and complex environments such as the Internet ecosystem (Iansiti and Richards 2006; Fransman 2014). We still have not reached a comprehensive operationalization of each of the three stages that characterize the EED approach in concrete routines. Further research should thus aim at identifying concrete procedures that allow for a full, detailed, and routinized exploration of each of these three stages. We consider that disciplines such as psychology, law, economics, sociology, engineering, financing, or ecology have a lot to contribute with to our further endeavors.

The book brings together two focuses of interests. On the one hand, based on an analogy from biology, it develops a new methodological approach, that has never been used to comprehensively explain the dynamics of the leading Internet business groups. It tries to combine an analysis that highlights the structure of these groups, primarily to provide an explanation of their evolution, and the consequences thereof. It could also help to better explain the evolution (foreseeable or otherwise) of the Internet universe. On the other hand, we expect that those readers, curious or interested in the present and in the future of the Internet, also find the book of their interest. We hope that the original explanation we pose to the dynamics, strategies, and diversification paths followed by some of the business groups covered in the book enlighten some of the readers (be them researchers or not), bringing them into a world of innovative thinking.

The introduction of continuous advancements in the Internet ecosystem has also posed several challenges for the authors of this book. When shall we stop following up the dynamics of the groups? Which of these should be included in the book and which should be left out? We are very much (and worriedly) aware that the rapid and disruptive evolution of the Internet and the groups leading it can make the book obsolete from the very moment in which it sees the light.

2 Summaries of Book Chapters

The chapters included in the book reflect and further elaborate on the issues discussed in this introduction. The book intends to balance both its conceptual contribution and the managerial implications that might be derived from the case studies included in it. With it, we aim to help extend the frontier of our understanding in areas where there are still significant gaps, both in theory and in practice.

Chapter "Introducing an Epigenetic Approach for the Study of Internet Industry Groups" by Mikel Gómez-Uranga, Jon Mikel Zabala-Iturriagagoitia, and Jon Barrutia starts elaborating by bringing into the floor concepts from molecular biology and business routines so as to build up the keystone concept in the book: Epigenetic Economic Dynamics (EED). The EED approach is defined as the study of the epigenetic dynamics generated as organizations adapt to major changes in their respective environments. Thus, its highest explanatory power is shown in rapidly changing environments, which entail fast organizational moves and/or decisions. The chapter is based on a biological analogy by which the concept of epigenetic is adapted and applied to the analysis of organizational dynamics so as to explore how organizations adapt to their respective environments in uncertain conditions as it is the case in turbulent and high velocity environments. The chapter discusses the analytical framework opened by the EED approach, which is implemented through a three-stage approach: (i) analysis of the environment and identification of the genomic instructions which are transmitted over time; (ii) identification of epigenetic dynamics; and (iii) identification and analysis of the consequences (in terms of innovation) as a result of epigenetic factors. The chapter concludes by providing some key characteristics for each of these stages.

Chapter "Epigenetic Economics Dynamics in the Internet Ecosystem" by Jon Mikel Zabala-Iturriagagoitia, Mikel Gómez-Uranga, Jon Barrutia, and Goio Etxebarria focuses on the analysis of the dynamics observed in the Internet ecosystem. In other words, the chapter explores in depth the second stage of the EED approach outlined in Chapter "Introducing an Epigenetic Approach for the Study of Internet Industry Groups." In particular, it analyzes the evolution and dynamics observed in some of the leading Internet firms, determining how these expand and diversify their activities in response to changes in their environments, in what we refer to in the book as epigenetic dynamics. The firms studied in the chapter are: Google, Apple, Facebook, Amazon, Samsung Electronics, Microsoft, Nokia, Twitter, eBay, and Yahoo. The chapter addresses the study of the dynamics of the previous firms through the analysis of their patenting behavior and the completed mergers and acquisitions. The chapter explains how the diversification paths followed by these firms go against all odds or predictability. The chapter finds how even if these firms have an original and dominant specialization (i.e., DNA), they are somehow forced to become active in areas that are unrelated to its original DNA, so that all groups get into the dynamics of the others in such a way that everyone competes with everybody on every possible vector of activity. Consequently, the chapter aims to find out to what extent the dynamics followed by the analyzed firms are more or less disruptive, and more toward or away from their original DNA. Finally, the chapter concludes by providing a brief analysis of the consequences of the previous epigenetic dynamics.

Chapter "GAFAnomy (Google, Amazon, Facebook, and Apple): The Big Four and the b-Ecosystem" by Juan Carlos Miguel and Miguel Ángel Casado focuses on four of the firms discussed in Chapter "Epigenetic Economics Dynamics in the

Internet Ecosystem" (i.e., Google, Apple, Facebook, and Amazon). It aims to show which factors contribute to explaining their quasi-monopolistic position and their influence on the development of the markets in which they operate. The chapter illustrates how the previous factors confer the firms' offensive and defensive powers, concluding that the firms behave as business ecosystems with related activities. The intention of the chapter is to explain the growth of these firms through their characterization as two-sided markets. The authors discuss how innovation is essential for the four companies, and illustrate why the stiffer the competition, the greater the pressure they face to innovate.

Chapter "The Digital Ecosystem: An "Inherit" Disruption for Developers?" by Jorge Vega, Jon Mikel Zabala-Iturriagagoitia, and José Antonio Camúñez-Ruiz aims to show developers' vision of the dynamics that occur in the Internet ecosystem. In this chapter, the authors take a look at the other side of the Internet, analyzing what is happening in the world of developers, what they do, and how the Internet is seen from the thousands of start-ups that form it. Do developers think that there is a disruption that is the same or equivalent to what the big Internet companies create in other scopes? That is the question addressed in this chapter. The chapter starts by illustrating the evolution of programming languages over the past decades so as to better understand the evolution of developers' logic and the different developer profiles. It does also discuss some of the possible future scenarios that might be expected, from the perspective of developers and start-ups and their role in the Internet ecosystem as a whole. The chapter concludes by discussing the potential utility and adequacy of the epigenetic (i.e., EED) approximation for studying developer-related dynamics in the previous scenarios and raises the possibility of their being studied from a quantum approach.

Chapter "Future Paths of Evolution in the Digital Ecosystem" by Jorge Pérez Martínez and Silvina Serrano Calle focuses on the technological progress and the expansion of the digital ecosystem. The authors discuss how the Internet has caused disruption in numerous scopes. It is not only a technological advance, but also an economic one, as it alters the conventional relationships established in the classic supply and demand model, and also what could be considered a more subversive element, that of human relationships, what has brought a new political and social order. As a result of the transition toward mobile devices, new challenges are being opened in relation to telecom providers and physical telecommunication infrastructures, which are evidenced and discussed. One of the key claims made by the chapter is the need for a global Internet governance. In this sense, the role of the Internet governance forum is highlighted as a body seeking to define a series of principles. The paper elaborates on the difficulties entailed in guaranteeing security and continuity of the net. Regulation of the net is an extremely complex task that goes beyond the virtual borders of each country's legal framework, and which has clear implications in terms of network security and control.

Chapter "4G Technology: The Role of Telecom Carriers" by Andrés Araujo and Itziar Urizar centers on the role of telecom carriers, analyzing operator strategies and performance in entering activities such as the entertainment industry, financial services or e-health. The chapter explores the evolution taken place in the development and deployment of the 4G long-term evolution technology, and the regulatory changes involved in such an evolution. It illustrates the large investments required in mobile technology which are driven by the whole of the ecosystem, operators, software producers, hardware producers and contents, services, and applications providers included. The authors tackle the regulation that has favored or in some cases hindered the development of the telecom sector and the digital economy in Europe in general and in Spain in particular. The chapter concludes by discussing how despite the large investments undertaken by mobile network operators, the growth in their revenues has stagnated in recent years, largely because their principal markets are saturated and because of competition from virtual operators and over the top firms also offering voice and data services using the networks deployed by the former.

Finally, Chapter "Scope and Limitations of the Epigenetic Analogy: An Application to the Digital World" by Jon Barrutia, Mikel Gómez-Uranga, and Jon Mikel Zabala-Iturriagagoitia concludes the book by addressing the scope and limitations of the epigenetic analogy. The chapter discusses in depth how economic analysis falls somewhat short of offering explanations and predictability of the digital ecosystem, concluding that an ecology-biology analogy may be appropriate as a body of knowledge to offer explanations and predictability. One of the problems encountered in any analysis undertaken within the Internet ecosystem is the definition of the borders. Setting topological markers proves difficult due to the dynamic nature of the ecosystem concerned, its powerful permeability and penetration in other production sectors and its eagerness to radically transform these. In the book, we have partially included the analysis of the consequences of epigenetic dynamics. However, since consequences are only observed ex-post, the undertaken analysis of is still rather partial, remaining as a matter of further work. There are new analytical challenges for the biological analogy in general, upon which the field of evolutionary economics is based, and for the EED approach introduced in this book in particular, that require the construction of models with explanatory and predictability potential so that they can orient policy makers' and business management's decision-making. This last aspect is an even more sensitive subject and calls for further research, which we aim, as a community, to be able to accomplish in the following years to come.

References

Aldrich, H. E., Hodgson, G. M., Hull, D. L., Knudsen, T., Mokyr, J., & Vanberg, V. J. (2008). In defence of generalized Darwinism. *Journal of Evolutionary Economics, 18*, 577–596.

Allekote, B., & Blumenröder, U. (2010). When patents become standard: Litigation for essential patents. *Building and enforcing intellectual property value*, 37–39.

Boschma, R. (2015). Towards an evolutionary perspective on regional resilience. *Regional Studies, 49*(5), 733–751.

Boschma, R. A., & Martin, R. (2007). Constructing an evolutionary economic geography. *Journal of Economic Geography, 7*, 537–548.

Carey, N. (2012). *The epigenetics revolution. How modern biology is rewriting our understanding of genetics, disease and inheritance*. New York: Columbia University Press.

Cetindamar, D., Phaal, R., & Probert, D. (2009). Understanding technology management as a dynamic capability: A framework for technology management activities. *Technovation, 29*, 237–246.

Corning, P. A. (2002). The re-emergence of "emergence": a venerable concept in search of a theory. *Complexity, 7*(6), 18–30.

Eisenhardt, K. M., & Martin, J. A. (2000). Dynamic capabilities: What are they? *Strategic Management Journal, 21*(10–11), 1105–1121.

Feldman, M. S., & Pentland, B. T. (2003). Reconceptualizing organizational routines as a source of flexibility and change. *Administrative Science Quarterly, 48*, 94–118.

Francis, R. C. (2011). *Epigenetics. How the environment shapes our genes*. New York: WW Norton & Company.

Fransman, M. (2014). *Models of innovation in global ICT firms: The emerging global innovation ecosystems*. Luxembourg: European Union.

Frenken, K., Van Oort, A., & Verburg, T. (2007). Related variety, unrelated variety and regional economic growth. *Regional Studies, 41*(5), 685–697.

Geels, F. (2002). Technological transitions as evolutionary reconfiguration processes: a multi-level perspective and a case-study. *Research Policy, 31*, 1257–1274.

Geels, F. (2014). Reconceptualising the co-evolution of firms-in-industries and their environments: Developing an inter-disciplinary Triple Embeddedness Framework. *Research Policy, 43*(2), 261–277.

Gómez-Uranga, M., Miguel, J. C., & Zabala-Iturriagagoitia, J. M. (2014). Epigenetic economic dynamics: The evolution of big internet business ecosystems, evidence for patents. *Technovation, 34*(3), 177–189.

Gottlieb, G. (2001). A developmental psychobiological systems view: early formulation and current status. In S. Oyama, P. E. Griffiths, & R. D. Gray (Eds.), *Cycles of contingency: Developmental systems and evolution*. Cambridge, MA: MIT Press.

Graham, S., & Vishnubhakat, S. (2013). Of smart phone wars and software patents. *Journal of Economic Perspectives, 27*(1), 67–86.

Griffiths, P. E., & Gray, R. D. (2005). Discussion: Three ways to misunderstand developmental systems theory. *Biology and Philosophy, 20*, 417–425.

Holland, J. H. (1998). *Emergence: From chaos to order*. Redwood City, CA: Addison-Wesley.

Iansiti, M., & Richards, G. L. (2006). The information technology ecosystem: Structure, health, and performance. *The Antitrust Bulletin, 51*(1), 77–110.

Jablonka, E., & Lamb, M. J. (2005). *Evolution in four dimensions: Genetic, epigenetic, and symbolic variation in the history of life*. Cambridge, MA: MIT Press Books.

Levinthal, D.A., & Marino, A. (2013). *Three facets of organisational adaptation: Selection, variety, and plasticity*. The Wharton School, University of Pennsylvania.

Markard, J., Raven, R., & Truffer, B. (2012). Sustainability transitions: An emerging field of research and its prospects. *Research Policy, 41*, 955–967.

Martin, R., & Sunley, P. (2007). Complexity thinking and evolutionary economic geography. *Journal of Economic Geography, 7*, 573–601.

Martin, R., & Sunley, P. (2015). Towards a developmental turn in evolutionary economic geography? *Regional Studies, 49*(5), 712–732.

Moore, J.F. (2005). Business ecosystems and the view from the firm. *The Antitrust Bulletin*. http://cyber.law.harvard.edu/blogs/gems/jim/MooreBusinessecosystemsandth.pdf. Accessed 1 Sep 2015.

Oyama, S., Griffiths, P. E., & Gray, R. D. (2001). *Cycles of contingency: Developmental systems and evolution*. Cambridge, MA: MIT Press.

Razavi, A.R., Krause, P.J., & Strømmen-Bakhtiar, A. (2010). From business ecosystems towards digital business ecosystems. In Proceedings of the 4th IEEE International Conference on Digital Ecosystems and Technologies (pp. 290–295), Dubai, April 13–16, 2010.

Robert, J. S., Hall, B. K., & Olson, W. M. (2001). Bridging the gap between developmental systems theory and evolutionary developmental biology. *BioEssays, 23*, 954–962.

Suárez, F. F., Grodal, S., & Gotsopoulos, A. (2015). Perfect timing? Dominant category, dominant design, and the window of opportunity for firm entry. *Strategic Management Journal, 36*, 437–448.

Teece, D. J. (2012). Dynamic capabilities: Routines versus entrepreneurial action. *Journal of Management Studies, 49*(8), 1395–1401.

Teece, D. J., Pisano, G., & Shuen, A. (1997). Dynamic capabilities and strategic management. *Strategic Management Journal, 18*(7), 509–533.

Tidd, J. (2001). Innovation management in context: Environment, organization and performance. *International Journal of Management Reviews, 3*(3), 169–183.

Witt, U. (2014). The future of evolutionary economics: Why the modalities of explanation matter. *Journal of Institutional Economics, 10*(4), 645–664.

Introducing an Epigenetic Approach for the Study of Internet Industry Groups

Miguel Gómez-Uranga, Jon Mikel Zabala-Iturriagagoitia and Jon Barrutia

1 Introduction

The aim of this first chapter is to study how organizations adapt to extremely fast qualitatively significant changes in the environment. As opposed to the prevalent Darwinian approach in which the logic of the phenotype is seen as a slow and moderate adaptation of social organizations to changes, our view focuses on rapid adaptation to quickly changing environments.

The analytical framework we put forth in this chapter, through the concept of Epigenetic Economic Dynamics (EED), comes from different fields of knowledge. This concept finds its roots in: (i) new discoveries in molecular biology; (ii) the complexity theory, which is a theoretical framework stemming from very diverse sciences; (iii) current approaches in terms of organizational routines in management; (iv) economic theory on competition and profits; and (v) innovation studies from a Schumpeterian approach.

Three related points could be cited as where to focus analyses concerning organizations' adaptation to changing environments: The mechanics of change in routines; the necessary capabilities that organizations require; and the resulting

M. Gómez-Uranga
Department of Applied Economics I, University of the Basque Country,
UPV/EHU, Bilbao, Spain
e-mail: miguel.gomez@ehu.es

J.M. Zabala-Iturriagagoitia (✉)
Deusto Business School, University of Deusto, Donostia-San Sebastian, Spain
e-mail: jmzabala@deusto.es

J. Barrutia
Department of Management and Business Economics,
University of the Basque Country, UPV/EHU, Bilbao, Spain
e-mail: jon.barrutia@ehu.es

© Springer International Publishing Switzerland 2016
M. Gómez-Uranga et al. (eds.), *Dynamics of Big Internet Industry Groups and Future Trends*, DOI 10.1007/978-3-319-31147-0_2

dynamics observed in them. The adaptation to changes in the environment in each case makes it possible to study these three approaches in a related manner.

This chapter will be particularly focused on molecular biology and business routines, respectively.[1] As Vosniadou and Ortony (1989: 1) discuss, "the ability to perceive similarities and analogies is one of the most fundamental aspects of human cognition. It is crucial for recognition, classification, and learning and it plays an important role in scientific discovery and creativity." The analogy from biology has been included in a great deal of the literature on organizational routines for over 50 years (Campbell 1965). Nelson and Winter's (1982) seminal work was a key reference for the dissemination of the evolutionary approach in economics (Witt 2008; Witt and Cordes 2007). Authors such as David (1994), Dosi (1982), Cordes (2006), Freeman (2002), Nelson (1995, 2007) and other post Schumpeterians use a Darwinian type of argument to defend their views. The principle of selection is a key part of Darwinian methodological approaches. According to it, organisms would gradually adapt in response to conditions determined by environmental factors. This analogy was greatly strengthened by the contribution concerning evolution of the genotype and phenotype and the latter's link with the 'plasticity' concept, which measures the degree of adaptation to change that defines the selection mechanism (Levinthal 1997; Levinthal and Marino 2013). The greater the plasticity, the lower the capacity to replicate, thus weakening the stability of routines. In contrast, without plasticity organizations would not have the ability to evolve and adapt so as to anticipate potential changes in the environment.

Paths of adaptation and learning have been explored for several decades in literature on management (Argote 1999; Levitt and March 1988). Parallel to this, new perspectives on dynamic capabilities have been developed (Teece et al. 1997), providing some methodological bases to advance in new directions. The complexity theory has been developed within the framework of organizational studies and strategic management. The interest of our conceptual contribution lies in understanding how organizations and businesses adapt to their environment in uncertain conditions. Organizations are hereby approached as complex adaptive systems, which show different principles such as self-organization, interdependence, co-evolution, complexity, and chaos. These principles form part of the development of our EED concept. Coevolution, as a case in point, means that entities, industries, or economies are partially linked to other organizations, or also that an organization changes according to the context (Kauffman et al. 1995).

Mitleton-Kelly develops (2003) and analyzes various perspectives on organizations and complex systems. She forms a theoretical framework that has been examined from very diverse sciences such as biology, chemistry, physics, mathematics, computation theory, economics, and the evolution of interactions, generally in ecosystems (Arthur 1999; Gleick 1987; Holland 1998; Petrosky and Prigogine 1990). As Mitleton-Kelly points out: "Although we make a conceptual

---

[1]Despite part of the current approaches to business routines being based on the complexity theory, we will not develop this point as it goes beyond the objectives set for this book.

distinction between a system and its environment, it is important to note that there is a dichotomy or hard boundary between the two, in the sense that a system is separate from and always adapts to a changing environment" (2003: 7). Kauffman (1993) suggests that natural selection is not the only source of order in organisms and that it is important to take self-organization into account because organisms also evidence spontaneous order, which is precisely self-organization.

Extrapolation of results from the past to the present becomes impossible when there is great complexity and dynamism in the environment. Reacting to changes in the environment through rapid flexible responses is sometimes the only solution. Many firms shift to an environment-driven orientation while others remain successful using their traditional formulas. Ansoff and Sullivan (1993: 1) "present a formula for strategic success which states that the profitability of a firm is optimized when its strategic behavior is aligned with its environment." In turbulent environments where there are large-scale changes, these can also be extremely rapid. "Thus, it is practically impossible to make predictions about the future, in which past experience would contribute little to adaptation. Even efficiently managed businesses will experience strategic surprises. In fact, the environment would change more quickly than possible responses and, in any case, strategic responses would seek new changes based on creativity" (ibid: 4).

In turbulent environments, where big changes are occurring, these adaptation processes need to be extremely rapid. Epigenetic changes form part of emerging processes. Emergence may take place as a result of significant changes in the environment and at greater or lesser speeds over time. It would be extremely useful in our model (i.e. EED) to be able to gauge the intensity and speed with which the successive epigenetic dynamics appear. Very rapid and turbulent changes in environments such as those that come from shifts in technoscientific paradigms have a considerable impact on business group dynamics (Gómez-Uranga et al. 2013). We think that these related facts are not always dealt with accurately and, in most cases, are not even envisaged in part of the literature that includes some type of biological analogy in its approach. This would be the case of the analysis of organizational routines or other advanced fields such as competitive models. One of our objectives when putting forth the EED concept is to explain situations found in real life and where the changes in the environment are extremely rapid and have a great impact. In this article we aim to contribute to better understanding of these dynamics.

The EED approach is understood as the study of the epigenetic dynamics generated as organizations adapt to major changes in their respective environments (Gómez-Uranga et al. 2014: 178). The concept shows its highest explanatory power in rapidly changing environments, which entail fast organizational moves and/or decisions. Some of the multiple causes of these changes include crises, changes in technoscientific paradigms, regulatory changes, massive acquisitions of intellectual property or other dynamic capabilities, strategic moves by competitors, etc. Insofar as these dynamics are disruptive, they can have economic (e.g. in terms of inefficiency), social, institutional, regulatory and even moral consequences.

Epigenetic dynamics are mainly due to economic rationality, which is also related to better innovation (in Schumpeter's sense of the term). Turbulent environments call for the adaptive capacity to act quickly. Thus, opening up new paths and achieving new objectives will also require participation from external bodies (business organizations or economic spaces) such as rapid actions to acquire and buy assets generated by other groups or in other places.

The chapter is structured as follows. The next section introduces the research gap addressed in the book, namely, how evolutionary economics cannot explain certain dynamics observed in high-velocity environments, and how new findings in biology, particularly those related to epigenetics, can help to bridge this gap. Section 3 provides an illustration of the complexity of the human genome, which serves as a starting point to introduce the concept of epigenetics and the advancement it provides to the understanding of evolution. Section 4 focuses on analysis of the organizational routines, and how these can also be studied through an epigenetic lens. Finally, Sect. 5 introduces the epigenetic economic dynamics (EED) approach, which is the cornerstone of the book, and which will be used to explain the dynamics observed in the Internet ecosystem.

2 Evolutionary Economics and the Research Gap

Since its beginnings, economics, as a scientific discipline, has imported knowledge which developed in other sciences (i.e. physics, ecology, biology, mathematics, etc.). In recent decades, biology has broadened its scope to penetrate a considerable part of the scientific production on economics.

Evolutionary economics finds its roots in a biological analogy, whereby economic systems behave like biological systems. Analogies are here understood as statements "about how objects, persons, or situations are similar in process or relationship to one another" (Van Gundy 1981: 45). The evolutionary approach toward the economy, following the principles of Darwinian theory, considers the changes that take place within economic systems as slow, gradual, and moderate. Thus, evolutionary economics is opposed to neoclassical economics, according to which economic systems are in situations of sustained equilibrium. In contrast, evolutionary economics would be in a state of continuous dynamism, although this dynamic may occur in a slow, gradual and progressive manner.

The main pillar of Darwinian principles centers on inheritance, where mechanisms such as replicas and descent act, and through which information concerning adoption is retained, preserved, transferred or copied over time (Darwin 1859, 1871). The principles of variance, selection and retention are a key part of Darwinian approaches. According to these, organisms would show a slow, moderate and progressive adaptation in response to conditions determined by environmental factors in order to survive, thus adopting different patterns and behaviors (Ansoff and Sullivan 1993). That is to say, adaptation is distinctive for being a slow, progressive and moderate process of evolution. However, as discussed by Gómez-Uranga et al. (2013, 2014), the interpretation of inheritance in an evolutionary classics framework is not the most suitable when trying to understand the evolution of the large Internet industry groups (see

Chapter "Epigenetic Economics Dynamics in the Internet Ecosystem" in this book by Zabala-Iturriagagoitia et al.). Just as human genome sequencing has, unfortunately, not completely explained the origin of modern illnesses, nor have evolutionary methodologies been able to decipher, and even less solve, the problems we encounter when interpreting the dynamics of Internet industry groups, which are the fastest growing on the world economy. Some of the defining characteristics of theories in the field of evolutionary economics are:

- The decisive role of the origins of each group as well as organizations' initial routes and DNA, or the 'first choice theories', which make it possible to explain later the paths they later follow.
- The evolution paths are almost charted, as is the case of: natural paths, the lock-in effect, replication, imitation, transmission of hereditary traits, selection and adaptation through gradual diffusion, etc.

Of the three key principles of Darwinism, variation, inheritance and selection, it is the latter that manages adaptive complexity (Hodgson and Knudsen 2006a, b). The selection principle shows why a group of self-organized units are able to survive by gradually adapting to their environment (Stoelhorst 2008). In business environments, this selection involves: conscious and deliberate choices, competitive pressure, market forces, environmental restrictions; all of which are put into practice through habits, routines, customs, technologies, institutions, regions, economies, etc. (Hodgson and Knudsen 2006a, b; Schubert 2012).

Geels (2014) has analyzed how evolutionary economics, neo-institutional theory and economic sociology conceptualize the co-evolution of firms and their environments, studying mechanisms of selection and adaptation and the tensions between them. In this regard, there has been a debate (particularly in Europe) on the extent to which analogy constructions using inputs from natural selection theory are useful in the evolutionary framework (Witt 2014). On the one hand, we find authors like Hodgson, who has influenced the development of evolutionary economics and evolutionary economic geography, and who introduced the concept of Generalized Darwinism. This concept follows the Darwinian analogy, although from a nondogmatic approach; i.e. with enough flexibility to be extended to various fields of the social sciences and economics. On the other hand, scholars such as Pelikan (2010, 2012) defend the need to transpose all genetic instructions, mechanisms and imprints into the biological analogy. From this latter perspective, the use of analogies would not be valid when concepts such as rules or routines are discussed unless they incorporate a series of precise instructions which could even be transposed into logical algorithms. Accordingly, Witt (2014) concludes that evolutionary economics today represents a patchwork of unconnected approaches. However, for evolutionary economics to be rethought, it first needs to include the general principles of new emerging fields, and then transpose them into concrete logics as suggested by Pelikan. Herstatt and Kalogerakis (2005) consider that analogies are based on surface and structural similarities. While surface similarities describe "the resemblance of target-objects to base-objects… structural similarities exist if relations between elements of the base object are similar to relations between various elements of the target object" (ibid: 333). In this regard, they consider that "the transfer of far analogies happens on a more abstract level than the transfer of near analogies and depends strongly on structural similarities" (ibid). One of these distant fields that might allow us to make an analogy for the purposes of this book is epigenetics.

The changes being perceived at the present time are characterized by their speed, constituting high-velocity markets and high-velocity environments (Eisenhardt and Martin 2000). It is in such environments that a number of dynamics are not being explained by evolutionary principles. As an illustration of the abrupt changes occurring in these high-velocity environments, the book focuses on the dynamics in the Internet ecosystem (Fransman 2014).

The principles that underpin evolutionary economics (i.e. path dependency, lock-in, replication, imitation, transmission of hereditary traits, selection and gradual adaptation) do not help to explain these fast dynamics. Evolutionary economics is thus unable to explain the dynamics observed in the Internet ecosystem (as an example of high-velocity markets) (Basole 2009; Jing and Xiong-Jian 2011). It therefore becomes necessary to redefine the principles of evolutionary economics in order to explain these fast changes that are increasingly occurring in most world economies. Addressing this failure and contributing to advancing the theory of evolutionary economics are the ultimate goals of this book.

This challenge (i.e. research gap) is relevant due to the systemic consequences that these dynamics are having on innovation systems (e.g. patent system, tax regulations, mobility of employees, training, etc.). It is also important to focus on this project at this particular time when the consequences of the previous dynamics are starting to be observed in multiple spheres worldwide (i.e. economic inefficiencies, blockage of competition, barriers to innovation, tax evasion).

As indicated, the book targets the literature on evolutionary economics, aiming to contribute to its further development by providing new parallels from biogenetics. Just as the origins of evolutionary economics go back to a (Darwinian) biological analogy, our positions come from that same starting point. Our reason for also supporting such an analogy is that, although the biological analogy allowed the development of the principles that laid the groundwork for the introduction of a theory on evolutionary economics, it has not been updated. The originality and ambition of the book lie in the fact that this stream of research has not been rethought in view of new findings from different fields within biology such as molecular biology, as discussed earlier, despite the origins of evolutionary economics, which go back to a biological analogy. That is to say, evolutionary economics is still governed by the same biological principles that were known in the 1970s and 1980s. To cover this research gap we will rely on the latest advances in the field of molecular biology, which, in recent years, have introduced the principle of epigenetics.[2]

---

[2]The biological analogy, following classical Darwinism, has been widespread for several decades. In fact, in the early 1990s some North American economists started a research stream around bioeconomics, from which journals such as the 'Journal of Bioeconomics' or the 'Journal of Evolutionary Economics' emerged. The biological analogy is, for example, dominant in some areas such as economic geography, within the evolutionary realm. Therefore, when introducing epigenetics, we do not require any legitimacy, as we are acting on an area where Darwinian biology has been widespread for decades.

The concept of epigenetics appeared some decades ago (Waddington 1953), but began to gain scientific relevance in the last decade of the 20th century (Francis 2011; Carey 2012). The reason for our focus on epigenetics, as social scientists with a focus on innovation studies, responds first to an inability to explain certain realities. This lack of a satisfactory explanation concerning certain realities is due to the fact that the biological analogy based on orthodox Darwinism is highly deterministic. In it, the gene determines the development of evolution. As stated, we observe that even if many expectations had been raised by the human genome sequencing, there is a certain amount of pessimism, because it has not lived up to the aspirations. And that is where epigenetics comes into the picture.

Epigenetics has shown that the DNA of organisms is not only susceptible to phenomena such as heredity, variety and selection (as derived from the Darwinian-type of biology). Instead, changes in the DNA of organisms can also be derived from (i.e. as a response to) changes in the environment. The environment in which an organism lives would therefore act as a 'traffic light' by activating or deactivating the expression of certain genes. Thus, the DNA of two organisms (e.g. twins), which is identical in origin, could evolve into different gene expressions depending on the environment they live in (e.g. whether one of the twins smokes, does sports, has different eating habits, lives in a city with high environmental pollution, experiences long-term unemployment that creates psychological stress, etc.).

Epigenetics allows us to update the previous evolutionary principles, as it is shown to be valid both in stable environments (i.e. slow, gradual and moderate changes) and in high-velocity markets and environments (i.e. fast, abrupt and unforeseen changes). In this book we focus on the latter type of environment. From the epigenetics perspective, we support the idea that adaptation of organisms/companies need not be gradual (Aldrich et al. 2008), and as will be illustrated, is sometimes very fast and even extremely abrupt.

In principle, epigenetics is linked to a different orientation of biology where molecular biology has become one of the dominant fields in recent years and new research fields are being opened up. We acknowledge that our initial point of departure is a heuristic approach. However, we have subsequently been able to create an ad hoc concept (i.e. EED) to explain these emerging realities in the Internet ecosystem. That is, we are able to develop an analytical framework and a methodological approach stemming from an initial conceptual dissatisfaction and which might be appropriate to explain the development and evolution of large business groups on the Internet that are becoming increasingly important and have dominated the world economy since 2004. The fast and intense development of these groups, places them at the forefront worldwide as per both market value and business profits (see Chapter "Epigenetic Economics Dynamics in the Internet Ecosystem" in this book).

The seminal work by Nelson and Winter (1982), provided the basis for the development of a microeconomic theory of organizational routines, based on a genetic analogy and following Darwinian principles. In addition, in recent years we find some approaches that may be close to what we observe in the field of epigenetics, namely, the development of a complexity theory in organizational studies (i.e. management), where organizations are treated as complex adaptive systems, and which present characteristics such as self-organization, interdependence, co-evolution, complexity and chaos (Holland 1998; Pohl 1999). Accordingly, our epigenetic approach is complemented or may be complemented by very different theories either related to the complexity theory or now being developed in other environmental sciences such as ecological systems (Folke et al. 2010).

With the EED approach we aim to identify the evolutionary dynamics of high-velocity environments such as the one found on the Internet. However, the key point of the definition of EED is that the adaptive changes are due to rapidly changing environments. Therefore, the decisions to adapt to these environments must be adjusted to the speed of the changes taking place in them. Some of the multiple causes of these changes, which should be identified, could include economic crises, changes in technoscientific paradigms, regulatory changes, values crises, massive acquisitions of intellectual property or other dynamic capabilities, strategic moves by competitors, etc.

Insofar as these dynamics are disruptive, they can have economic (e.g. in terms of inefficiency), social, institutional, regulatory and even moral consequences. In-depth analysis of these consequences is still needed. So far, in a previous research we have only pointed out some of the potential consequences of these dynamics for the patent system (see Gómez-Uranga et al. 2014). However, analysis of the previous type of consequences also needs to be addressed. An example currently under discussion in the European arena, are the increasing tax engineering practices of the large Internet business groups, which are having an increasingly greater influence in more European countries (Corkery et al. 2015; Heckemeyer et al. 2014; Li 2014). In this regard, the European Union is trying to stop or mitigate these tax engineering practices, which are affecting their respective member states.

In a recent article, Martin and Sunley (2014) intend to build a new framework for evolutionary economic geography. These authors believe that both the seminal work by Nelson and Winter (1982) and the studies by Boschma and Frenken (2006, 2009) see genes as the main replicators of biological information and recognize them as equivalent to business routines. Witt (2008) is of the opinion that the metaphors of conventional Darwinism cannot understand human creativity and learning. In turn, Nelson (2005) differentiates between sociocultural and biological perspectives.

Although the authors cited above do not adhere to some of the postulates proposed by Generalized Darwinism, neither do they reject them completely. From the perspective of the EED, the genome can be modified by changes in the environment (i.e. changes in the environment can lead to gene regulation causing them to express themselves or not), which would mean that there would not be a sole deterministic genetic inheritance. This is not envisaged in Generalized Darwinism. What they do propose is a wider array of concepts through a new methodological space that ideates two complementary research areas: Evolutionary Developmental Biology (EDB) and Developmental Systems Theory (DST).

Developmental Biology is concerned with ontogeny, with "the origin and development of an individual organism through its life span" (Martin and Sunley 2014: 717). One of the seminal works on Developmental Systems Theory (DST) is by Gottlieb (2001), who discusses the developmental systems view or probabilistic epigenesis. Another source is the DST put forth by Lewontin (1982, 1983), who criticizes the "lock and key model" approach in which organisms follow evolution adapted to the predetermined niche for which they were conceived and cannot leave. As we shall see in the following sections, an analogy could be drawn with the deterministic approach of path dependency (Martin 2012b).

The DST "stresses the delicate dependence (contingency) of development on a rich matrix of factors outside the genome" (Griffiths 1996; Griffiths and Gray 2005: 419). Advocates of DST criticize what they call genocentrism and align against those theories that understand evolution as revolving around DNA accompanied by some classic selection mechanisms. The authors place greater importance on factors other than DNA and defend the multiple factors of epigenetic inheritance. The properties of robustness and plasticity are defined in the analytical framework of DST. The first of them could be translated as the capacity to adapt to disturbances while maintaining system-specific functions (Kitano 2004).

Martin and Sunley (2014) use a metaphor for DST, stating that genes should be contextualized, as the environment sets new emerging (genetic) realities. Self-organization and emergence are the core DST properties most highlighted by these authors. As regards self-organization, the system components themselves change as a result of the activities they undertake. Interaction between system components leads them to greater complexity without there being any previous detailed instructions to follow. From our point of view, there are considerable parallels between the properties of emergence (Martin and Sunley 2014), adaptability (Boschma 2015) and plasticity, which we introduced in the previous section, and with successive epigenetic type processes, which form a logical base for our EED model.

On the other hand, the above authors highlight development as an "emergence process." Systems are formed at less-complex interaction levels in these development and emergence processes. The biggest generator of emerging evolutionary innovation is the environment (Oyama et al. 2001; Robert et al. 2001). Emergence is understood as a source of innovation and a dynamic process. New products and new firms in economic development systems can emerge as externalities from spatial agglomeration in the systems (Martin and Sunley 2007). A local cluster might shape a broader field or industry which it joins and then acquire an external reputation, which would in turn influence its own resources and lead to a better market position. Put differently, firms influence the environment and it therefore impacts local firms. Following this logic, path dependency could be understood as a type of emergence similar to the one we observed in some of the examples where we compared the dynamics of certain business groups studied in Gómez-Uranga, et al. (2014).

Evolutionary Developmental Biology (EDB) coincides with DST from another perspective. EDB goes beyond what is called Neo-Darwinist synthetic theory. EDB (dubbed Evo-devo) is based on advances in molecular genetic biology and envisages new genes being created from parts of old ones. It includes the authentic logic of epigenetics, which is gene regulation. It has been shown that evolution alters developmental processes to create new and original structures from the old gene networks. The differences between species are not found so much in the genes as in their expression (i.e. genetic switch).

Both approaches, therefore, coincide on the role of the ecological development context, which includes non-DNA factors and exerts a causal influence on gene expression (Gilbert 2001). That environment is the product of evolution (Griffiths and Gray 2005). This is the exact definition of epigenetics which inspires the concept of EED we use in this text.

EDB also envisages the property of plasticity and shows that phenotypes are not necessarily determined by their genotypes (West-Eberhard 2003). The existing morphological variety is not always reflected in the genome. Epigenetic dynamics are formed as mechanisms for evolutionary innovation.

Complex Adaptive Systems (CAS) focus on the instruments and agents that participate in decision-making in complex situations. In this case, the analytical tools (above all, mathematics) must adjust to nonlinear and nondeterministic processes (Holland 1998) as opposed to rationalism capable of mathematical predictions, operating on a previous order of subsystems which form an integral system in an organized manner.

Complex systems work as a network of different groups that act on identical time coordinates in order for each one to adapt to its environment (e.g., these may be individuals, social agents, firms, governments, etc.) (Pohl 1999). They adapt to their experience in the system and find themselves subject to the laws of natural selection (Holland 1998).

The variety of institutional frameworks that affect a system (rules, routines, habits, etc.), and the interaction between agents indicate its complexity. For several decades, there has been a great deal of literature on national, regional, and local systems where innovation plays a key role (Lundvall 1992; Cooke et al. 1997; etc.). These could also be understood as complex systems.

In this chapter, it is important to point out that the universe of complexity connects as a logical derivative with the property of unpredictability, which coincides with our EED methodological approach. CAS evolution mechanisms are related to low-level interactions between different agents in the system. Control and steering of the systems should focus on a decentralization logic, as opposed to centralization, monopolistic, and/or hierarchical structures. All in all, interaction in networks comprises key properties for systems that are adapting over time, all of which would be analogous to natural selection mechanisms (Pohl 1999). According to our vision, these CAS models (particularly concerning control and leadership) would not serve for adaptation to very dynamic environments.

We find methodological bases in the chaos theory to evaluate some of the areas we are most interested in exploring. For instance, the chaos theory sheds light on decision-making with its vision of how reality flows, seeing it not as exclusively chance occurrence (i.e. probability) but neither as a purely deterministic result. Neoclassical economics imitates classical physics, arguing in favor of stability (negative feedback), whereas the chaos theory envisages change and instability (positive feedback). Positive feedback establishes covariance relationships: when one variable increases, so does the other.

The analogy of the climate (i.e. environmental) is widely used in the social sciences. Organizations and territorial systems have predictable and unpredictable behaviors and it is not possible to discover all the factors affecting them, as sustained by Godel's incompleteness theorem. In contrast to behavior in terms of balance, organizations are out of balance, in situations that lead to change, in other words, to new imbalances and so forth.

We have already raised one of science's key issues, which is "reality." The complexity paradigm revolves, above all, around how we perceive reality rather than reality as objective information. Therefore, fractal geometry is used. Fractals are used to study nonlinear dynamic systems, and are also very useful to study decision-making in uncertain situations, for example, situations that shift very rapidly from stability to severe disturbance.

Decision-making agents are not individual actors. There is a social decision apparatus (Lara 1991) which is, in essence, a group of individuals who have the responsibility of high impact decision-making (for the future of an organization or territory). They should be capable of anticipating the behavior of the environment (Hernández-Martínez 2006).

Time in organizations can be seen from two perspectives: The first is irreversible and the second may be reversible (Etkin and Schvarstein 1995). Strategic decisions center on achieving the organization's medium and long-term objectives. Many of the variables that involve organizations are beyond their control and are therefore closely related to the time required to adapt to the external environment. In contrast, others take the organization's own internal dynamics or those of the actors in the different systems they belong to as their references.

In terms of external time, the response has to be agile to address the competition's actions and possible rapid market developments. In terms of internal time, however, this should not be considered long or short but simply synchronized with the organization's other cycles or those of the system concerned. Time cycles can accordingly be situated in different dimensions. For instance, we would not be in the same time space if we compared a software company, where the pace of technological development forces rapid strategy development, with a higher education institute where the cycles of current scientific knowledge are longer and strategies can therefore cover longer time periods.

Time is a key element in any analysis of adaptation, and thus of evolution. In our analytical framework (i.e. EED), we think a fast competitive response (from an economic viewpoint), and using exclusively the capabilities provided by the internal (or own) organizational/geographic framework would only be possible if the assets obtained through a significant efficient adaptive process were available at that moment, following a certain path dependency. According to our logic, two types of reactions with different characteristics would be observed (Table 1).

We would also like to underscore the differences between organizations' and territories' decision-making in extremely dynamic environments and in other more stable environments. In the first case, decisions should be made in very short time periods to achieve good adaptation and adaptive capacities should therefore be available either internally through possible acquisitions or as the result of cooperation with external agents established beforehand (i.e., not as a reaction to the situation). In rapid adaptation, the most efficient results could be achieved when

Table 1 The timing of adaptation in complex systems

(Radical) adaptation to the environment in the short term	Gradual adaptation over time
Faster and breaking with the past	Slower and extended over time
Relatively short time periods	Longer time periods
Greater previous adaptability	Less adaptability a priori, routi-nized operations
Adaptive dynamic capabilities	
Top management plays a key role in policy-making	Economic agent systems structured around path dependency
There is indirect influence from a wide variety of agents	The role of stakeholders
Decision-making authority concentrated in few hands	Very diverse agents
Decisions based on simple norms, heuristics, and specific programs	Decisions based on governance over time
Action based on structured information (i.e. reports on the economic environment, above all, competitors), and some actions improvised in a very short time	More formalized (routinized) actions
	Heterogeneous decisions over time
	Previous designs of Porterian clusters

Source Own elaboration

strategic decision-making is more concentrated, which would enable management to base it on simple rules and heuristics (Bingham and Eisenhardt 2011, 2014). A group of actors may influence management/policy makers. However, they do not play a direct role in strategic decision-making. Conversely, in more gradual adaptation processes that are sustained over time (i.e. they occur in stable environments) decision-making can be more routinized. That is to say, it can involve a wide variety of agents that form the system (i.e., complex systems), thus meaning that governance (either at the business or territorial level) is considered to be important. However, in spite of the inherent complexity of these multiagent systems, decisions are normally made by a small group of agents or individuals (e.g., CEOs and managers at the company level, ministers, etc. at the territorial level).

If a gene is not regulated (because the environment does not act on it) time may tend to slow down. In contrast, if the context (the environment) makes it regulate, time could be accelerated (a quicker intervention would be needed over time). That is to say, if certain institutions (routine, habits, rules, etc.) are maintained without the environment inducing an institutional change, time may go more slowly and even stop, until the moment when changes become inevitable because the organization or territory's very existence is being threatened.

Measuring time (objectively) is not the most significant point in our case. For example, the time shown in Fig. 1 that elapses between t_0 and t_1 is the same in both cases. However, the content of the changes from t_0 to t_1 is completely different because there are changes in both the environment and its intrinsic characteristics. We are not as interested in measurable time (t_0 to t_1), which is the same in both cases, but focus on the changes that occur in the environment during that same time interval.

The time elapsed would be the same in the two cases. However, in the second case, the amount and scope of the changes are much greater. In this case, the short, medium, or long terms do not mark measurable time, but the changes that are

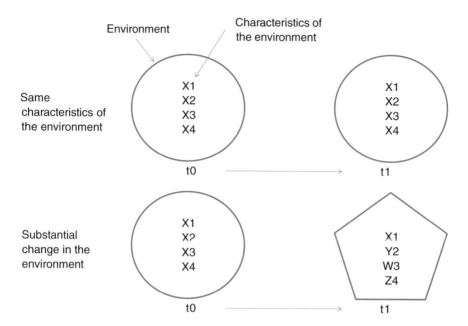

Fig. 1 Time and substantial change in the environment. *Source* Own elaboration

taking place. Therefore, the dynamics would not be studied by measuring time (in normal time units) but by analyzing the changes that occurred. In that case, the responses cannot be dated over time, but would occur according to the intensity and extent of the changes, such as for instance, significant intense changes in the environmental factors. We could calculate what changes in the environment (the arrangement and volume of its factors) call for reactions and also find out what level of reactions.

When we say an environment changes very quickly, we mean that we have a perception, a memory of the changes being faster than at other times when we perceived them to be more normal. For instance, in crises there are dynamics that move much faster than during noncrisis periods. If the pressure of competition from business groups increases, decision makers may have the perception that they are pressed for time and have to react or face changes as quickly as possible and should therefore devote exceptional effort to making time periods shorter. On the contrary, if there were no pressure from the competition, time would seem to pass much more slowly even if the measurable time were the same in both cases.[3]

---

[3]If, for example, we compared the changes that will occur in Internet industry group environments with the recent past, perhaps we should not talk about very fast changes because the current speed of change has become 'quite normal'.

Complexity can only be compatible with the irreversible (and often unpredictable) nature of processes, and this last concept is in some manner the opposite of resilience understood in a more mechanistic sense, used, as we have seen previously, in certain models of thinking in disciplines such as engineering or economic geography.[4]

Evolutionary approaches point to a trade-off between adaptation and adaptability. Adaptation concerns those changes within preconceived paths (Boschma 2015: 4). In turn, adaptability would respond to dynamic capabilities with multiple potential paths that evolve and place the local economy in a better position to face up to possible unforeseen developments. There would therefore be two different types of resilience, one that shows adaptation of previously existing conditions (new paths, inertia, and weak attachments between the place's social agents, etc.) in the short term. The other would be a response that would have to be formulated in the long term, involving certain breaks from existing conditions (new paths, weak inertia, and weak links between agents).[5]

The following can be cited as authentic examples of different types: system shocks such as natural disasters in certain geographical areas, global economic crises, technologies becoming obsolete, and slow burns or long-term movements such as reindustrialization, modernization, urban renewal, political transformations, etc. (Pike et al. 2010; Pendall et al. 2010; Boschma 2015; Martin 2012a). In our approach (EED), external shocks are not only those resulting from changes in natural environments, or even those macroeconomic problems caused by a crisis. They also come from competition, mergers, and massive patent acquisitions, changes in technological paradigms, changes in demand, etc. which are more similar to the dynamics of business groups (Gómez-Uranga et al. 2014). Long-term strategies call for changes and innovations, institutional changes, as well as strategies to destroy old ones and allow the creation of new paths (Martin and Sunley 2007) to address changes in the environment.

Adaptability is more closely associated with the concept of plasticity (Levinthal and Marino 2013). From our point of view, literature on evolutionary economic geography is taking a mechanistic approach when presenting the concepts of adaptation and adaptability. In our opinion, adaptability is a property that serves to adjust to changes in the environment. However, without adaptability there cannot be adaptation in the short or long term.

---

[4]From another perspective, the second law of thermodynamics has one main purpose: Impose a strict world symmetry on the directions of the time axes towards the past and the future (Davies 2002). That is to say, there is asymmetry in both directions, past and future. And the essence of this asymmetry lies in the changes that have taken place. As Davies states (2002: 10) we do not really observe the passage of time, but we are actually observing how the later states of the world differ from earlier states that we still remember. A watch does not really measure the speed with which one event follows another. Therefore, it appears that the flow of time is subjective, not objective (ibid: 11).

[5]Folke et al. (2010) develop a concept of resilience, which we believe has a greater scope than the one used by other authors. In their approach, resilience is formulated as the capacity that a complex 'social ecological system' has to continuously adapt, which is a more reasonable and less 'mechanistic vision than others used for the concept, as they are mainly limited to the field of economic geography.

The more diversified economies are, the more adaptable they will be in the long run, and therefore, the greater their capacity to adapt to new growth paths (Pike et al. 2010; Frenken and Boschma 2007). In this sense, extensive empirical evidence surrounding the concept of related variety concludes that the industrial path (i.e., history) of each territory is key to understanding the new adaptation processes that are going to occur in them (Boschma 2015).

At certain times and in certain situations, particularly in a globalized economy where related varieties are global, industrial activities can also take place outside the region. In other words, this is not only an endogenous approach. If there is an external shock, one fast way to adapt is, for instance, to buy patents, form alliances, mergers, joint ventures, etc. with agents from outside the region. That is to say, the bases needed to establish related variety may not only come from within the territory. In a global economy, we have to consider that related variety is also generated in industries and agents located outside it, which means that we have to resort to industries/agents located beyond the territory. Furthermore, spin-offs stemming from ex novo relationships with other agents such as universities, technology centers, etc. which have very little to do with the territory's history could appear (i.e. unrelated variety). According to Boschma (2015: 9), unrelated variety would guarantee adaptability, while related variety would secure adaptation. Therefore, having both types of variety would make a territory (i.e., an organization) truly resilient.

In biological complexity, variety also means that not everything comes from some initial origins (i.e., such as genes). Total path dependency does not exist as there is a part which stems from the environment. A part of regional development clearly comes from each territory's history, the existing values, its institutions, etc. However, there are activities which lead to dead ends in spite of having a historical base (e.g., appliances in historical industrial regions). Or activities having an ex novo nature take place, which do not come from any previously existing relationship or related activity that may have existed in the region prior to that time.

Geographers such as Pike (2002) base their idea of adaptation on previously existing paths. The focus would be on adapting to major turbulences such as emergencies or disasters. Authors like Teece (2007) believe that ordinary or previously existing capabilities are due to routinized behavior. While geographers view adaptability as a systemic property responding to slow changes taking shape in the long term, Teece's dynamic capabilities are based on changes and adaptation to fast changes in the environment, a view which has also been shared and reinforced by Eisenhardt and Martin (2000).

Environmental changes are the ones that mark time for organizations because they must deploy the most suitable adaptation dynamics. Going one step farther in our argument, in order to put it into practical terms for system agents, it is necessary to know the time (measurable) that would be needed for environmental changes to take place. However, since prediction is impossible, in any case, comparison can be made between the time and/or frequency that they appear with the frequency of other similar changes that we have known or remember. This is what occurs with predictions about scientific progress in the future (e.g., shifts in scientific-technological paradigms). In this way, we could classify them as fast, very fast or slow. It is much more difficult to predict when those changes in the environment will happen than to foresee what the response should be (for example, from organizations) when they occur.

3 Epigenetics Beyond Darwinism: The Complexity of the Genome and (Human) Life, and Importing These Concepts to the Social Sciences

Life is a major source of complexity, and evolution is the process describing the increase in this complexity. In other words, evolution leads to higher complexity, and complexity and emergence are two interrelated processes. As discussed by Corning (2002: 27), *"in evolutionary processes, causation is iterative; effects are also causes. And this is equally true of the synergistic effects produced by emergent systems. In other words, emergence itself… has been the underlying cause of the evolution of emergent phenomena in biological evolution; it is the synergies produced by organized systems that are the key… a change in any one of the parts may affect the synergies produced by the whole, for better or worse. A mutation associated with a particular trait might become "the difference that makes a difference"…, but the parts are interdependent and must ultimately work together as a team. That is the very definition of a biological whole."*

To introduce the EED approach in greater detail, we need to go back slightly to the state-of-the-art in biology and then move beyond classical and orthodox Darwinism. The most recent results found in molecular biology have shown that genetic structure is highly diverse and complex. As an example, the human genome only contains a small number (approximately 2 % of the total) of modifying genes that encode (transfer the hereditary information/instructions), the proteins. There are also other noncoding genes (RNA) which act on the coding genes as "messengers." In recent years, it has been discovered that RNA also has other functions that have not yet been clearly defined. RNA is known to play a role in regulation, which is carried out jointly with other nongenetic elements (ENCODE Project).[6] There is also another part of coding DNA whose function is still not clearly understood to date.

Some years ago, it was believed that once the human genome was sequenced, it would provide us with a map to decipher/interpret everything that could happen during a person's life. However, this has failed, at least in part. The geneticist and philosopher Ayala (2013) proposed the following analogy with computers to explain life: the information on how to build the computer is also contained in the computer itself. In other words, it needs both parts, hard (to process the information) and soft (the information itself). This metaphor illustrates the complex meaning of life.

---

[6]See https://www.encodeproject.org/ (last access October 2015).

Gene expression is highly regulated, thus enabling it to develop multiple phenotypes (Masuelli and Marfil 2011) that characterize the different cell types in an organism, thus providing cells with the elasticity to adapt to a changing environment. In other words, genes can be expressed or not expressed in terms of interactions, depending on how these occur with the environments (Lewontin 1982, 1983). Changes in the environment may cause chemical changes that affect certain proteins (histones). Depending on the conditions, they may alter gene expression, activate or deactivate coding genes and their expression (i.e., like a 'traffic light'). These are called epigenetic processes (Carey 2012).

Waddington coined the term epigenetics in 1953 to refer to the study of interactions between genes and environment that take place in organisms. Epigenetics centers on knowing how, when, and why gene expression is regulated. In developmental genetics, epigenetics refers to the gene regulation mechanisms which do not involve changes in DNA sequences, but are still passed down to other generations (Francis 2011). It mainly focuses on understanding the influence of the environment on genome expression; in other words, changes in gene expression that can also be transmitted and inherited (Canetti 2003). One of the key sources of gene modification is the environment, and it can affect one or several genes which carry out multiple functions (Carey 2012). Epigenetic regulation shows how the plasticity of the genome enables it to adapt to the environment, resulting in the formation of different phenotypes determined by the environment that the organism is exposed to (Cavagnari 2012; García Azkonobieta 2005; Waddington 1947, 1953).

According to Evolutionary Developmental Biology (Evo-Devo), the morphological variety shown in the various *"clodes"*, is not always present in the genome but is also caused *by* mutation-driven changes in gene regulation.[7] Biodiversity is often not brought about by differences in genes but by gene regulation (epigenetic changes) (Carroll 2005).

Epigenetics is understood as changes in gene expression that are transmitted to cell division and sometimes between generations but do not involve changes in the underlying DNA sequence (which was the mainstream belief in twentieth century evolutionary science).[8] The epigenome enables a relatively rapid adaptation to the environment, without the change being recorded in the genome (Weitzman 2011). The phenotype is determined by the activity of many enzymes and their interaction with proteins. Thus, changes in the environment (e.g., temperature, pollution) can lead to changes in the phenotype.

Epigenetics leads to abnormalities and changes in what is programmed or initially encoded. Changes in external conditions determine the different ways habits and routines are expressed. Business routines should go hand in hand with circumstances or features needed for them to be expressed properly. A change or modification in these accompanying features may lead to changes in the normal

---

[7]The capacity to transmit epigenetic marking between generations translates as chemical changes in the chromatin structure, which may be greatly determined by environmental factors.

[8]See http://www.epigenesys.eu (last access October 2015).

expression of those business routines. For instance, they may be translated as inexplicable behavior or practices (i.e., an illogical result) of the initial information transmitted from generation to generation (i.e., as if they were mutations).

In our conceptual business analogy, evolution is much faster than in biology (Abatecola et al. 2015). Genes mutate or change in much shorter time periods. We run into a different time dimension. The characteristics that identify business groups' genomic instructions evolve over time so that they are sometimes a mere enlargement of previous functions and at other times are more radical changes. However, they always maintain a thread connecting them to the business's initial specialist field.

According to Mortara and Minshall (2011: 591) these "revolutionary changes" are needed in certain industries due to the speed at which changes occur in their high-velocity, turbulent, and unstable environments (Eisenhardt and Martin 2000; Suárez 2014). The dynamics that we refer to as epigenetic cannot be interpreted or foreseen from organizations' initial competences, activities, resources, and routines. Above all, epigenetic dynamics respond to an economic rationality, which is also linked to the development of innovation (in Schumpeter's sense of the term).

In the biology of species, individuals have to behave differently when faced with the need to compete or defend themselves. This synergistic adaptation to the environment may produce a more favorable phenotype (Gilbert and Epel 2009). The complexity of life (particularly human life) cannot be translated or treated merely as genetic code sequencing: the human genome provides necessary information for a coordinated regulated expression of the genetic makeup. The set of "expressed" (i.e., active) proteins (i.e., the proteome) carries out most of the cell functions (e.g., enzymatic, metabolic, and regulating) through an enormous amount of practical networks. These are the cells' structural makeup that forms the tissues and organs of living beings.[9]

Table 2 provides some simple features of epigenetics. The bases of the genome are more complex than expressed in orthodox Darwinisim, as there are some knowledge gaps concerning the exact functions of the different elements that make up the genetic structure (e.g. RNA, nongenic bases, genetic garbage, etc.). In recent years, it has been observed that these parts, which did not a priori play a role in bringing such information to proteins for the development of life, play a more important role than previously expected. Our main point of interest lies in classic selection processes. It is here that we find the part of both biology and genetics that expresses the gradual and moderate adaptation of species, which adapt to changes that could lead them to become more efficient. These changes may contribute to the genetic heritage, together with the acquired characteristics, mainly in stable environments. In turn, epigenetic processes occur through changes in unstable or turbulent environments. The EED approach is highly suited to explain the adaptation to the latter type of contexts, where the contribution to the genetic heritage would occur together with the acquired characteristics as a result of these adaptation processes to very rapidly changing environments.

Another property of epigenetics that must be taken into account before we move forward to conceptualize our EED approach is the underlying uncertainty about the

---

[9]In this sense, human cells resort to splicing, producing several proteins with very different functions from the same gene (ENCODE Project).

Table 2 Some features necessary to understanding epigenetic dynamics

Bases of the genome	Phenotype and epigenome	Results
Complexity	Classic selection processes through gradual moderate adaptation	Contribution from genetic inheritance jointly with features acquired as a result of (gradual) adaptive processes to (stable) environments
	An epigenetic process through rapid changes to adapt to turbulent environments	Contribution from genetic inheritance jointly with features acquired as a result of (rapid) adaptive processes to (unstable) environments
Different RNA functions		Human influence in epigenetic changes themselves
Knowledge gap concerning the precise functions of the different elements that make up the genetic structure (RNA, nongenetic, etc.)		Uncertainty concerning the results of the processes

Sources Own elaboration

results of these processes. When we draw the analogy to bring the concept of epigenetics into the study of the dynamics of Internet business groups, the main feature is the inability to forecast the possible or potential dynamics that will characterize the evolution of these groups (see Chapter "Epigenetic Economics Dynamics in the Internet Ecosystem"). Our goal is to find an analytic framework that allows us to better understand that these epigenetic dynamics are not marginal, but rather voluntarily sought and thus, are due to the economic rationality of these business groups, which is evidenced by the new paths, activities, and industries they move into, following different strategies. However, given the fact that the prognosis of their evolution is very difficult, it is possible to refer to this industry as a highly uncertain environment.

Going back to orthodox Darwinism, many authors have imported Darwinian principles of biology to fields and methodologies in the social sciences such as evolutionary economics or evolutionary economic geography (Breslin 2011; Aldrich et al. 2008; Boschma and Martin 2007, 2010; Essletzbichler and Rigby 2010; Hodgson and Knudsen 2004, 2006a, b, 2012; Pelikan 2010, 2012). Hodgson (1993, 2009, 2010, 2012), who has been influential in fields such as evolutionary economics or evolutionary geography, can particularly be cited as one of the authors who most centered on transferring these principles from biology to the social sciences. Hodgson, in line with other scholars, introduced a more flexible approach than orthodox Darwinism and called it Generalized Darwinism (Mayr 1988, 1991; Aldrich et al. 2008; Hodgson and Knudsen 2006a, b, 2012; Levit et al. 2011), which takes a non-dogmatic approach to the Darwinian analogy (i.e., namely, with enough flexibility to be extended to various fields of the social sciences and economics).[10]

---

[10]"Given that the entities and processes involved are very different; these common principles will be highly abstract particular domain. For example…, we can generalize principles that apply to all the phenomena, despite major differences in their features. In biology and in the social sciences, the phenomena are so complex that scientists supplement general principles by many more auxiliary and particularistic explanations, thus differentiating these sciences from physics" (Aldrich et al. 2008: 580).

We agree with part of Aldrich et al.'s (2008: 578) arguments in defense of this concept when they observe that the principle of selection "could help explain survival not only for individuals, but also of groups, customs, nations, business firms and other social institutions."

> There must be an explanation for how useful information concerning solutions to particular adaptive problems is retained and passed on. This requirement follows directly from our assumptions concerning the broad nature of complex population systems, wherein there must be some mechanism by which adaptive solutions are copied and passed on. In biology, these mechanisms often involve genes and DNA. In social evolution, we may include the replication of habits, customs, rules and routines, all of which may carry solutions to adaptive problems (Aldrich et al. 2008: 584).

However, from our point of view, proponents of Generalized Darwinism encounter a number of difficulties to adapt this approach to a relevant share of the changes that occur more and more rapidly, particularly in times of crisis, and which are becoming more important and having bigger impacts on a number of dimensions (i.e. social, technological, economic, institutional, moral, etc.). The central argument in Generalized Darwinism continues to be based on a phenotypic selection, and therefore differs only slightly from an orthodox conception of Darwinism.

Darwinism considers that organisms gradually adapt in response to conditions determined by environmental factors. This interpretation of inheritance followed by the classical evolutionary framework is not the most suitable when trying to understand the evolution of large Internet industry groups, which are characterized by their sudden and radical dynamics (Deighton and Kornfeld 2013).

For Hodgson and Knudsen (2012), replicators are the basis for genetic inheritance and are specifically found in processes such as the transfer of rules, norms, and business routines. The authors distinguish between replicators and interactors.[11] From our perspective, the epigenetic analogy could go farther if we consider that the initial bases which are to be replicated can be expressed in different ways (Gillham 2001). In other words, this would not be a selection or adaptation process of the immutable inherited base in its strictest sense, as is understood in the most widely accepted and frequent interpretations in biology.

As they advanced in their studies, Hodgson and Knudsen proposed understanding Lamarckism "as the inheritance of acquired characters" (2012: 14) when identifying the social replicators (genotypes) and social interactors (phenotypes). The same authors went on to state that *"in order to consider and understand the possibility of Lamarckian inheritance we must first identify the replicators and interactors in the social domain. We must then consider that the acquired character of an interactor can affect its replicators… Further examples of social replicators include routines, by which we refer to dispositions within organizations to carry out sequences of actions. Routines are hosted by organizations as their interactors, and in turn are built on the habits of the individuals involved"* (ibid: 16).

---

[11]The main controversy between Hodgson and Knudsen (2012), Pelikan (2010, 2012) and Levit et al. (2011) centers on the role that replicators play in evolution.

We believe Lamarck's approach still proves useful as a means to go beyond the orthodox lock-in. Hodgson and Knudsen's contributions when searching for a type of Lamarckism that could be useful in the social field are also interesting. However, in view of what we are observing in this book, we think it is logical to follow a more direct path. In order to do so, we make use of current contributions from the sciences: genomics, proteomics, etc. Moreover, we focus on a more flexible concept of organizational routines than what we find in today's literature on management.

4 Routines: Complexity and Adaptation

In our methodological approach, routines or replicators are flexible and complex. We observe an analogy with the latest findings from molecular biology which we mentioned in the previous section. Certain routines are replicated and others are not. Instructions can be transmitted exactly as they are formulated while others disappear and give way to new ones. Transmission is neither simple nor automatic and, on certain occasions, elements that are not in the body of instructions are transmitted.

Recent studies on routines, in the field of phenotypes, deal with the mechanics of change in the internal and external routines of organizations to enable them to adapt to rapid changes in the environment. Changes in routines, as well as the plasticity needed to readjust routines on a permanent basis, depend on the capacity to reconstruct and reorganize resources and competences (in Teece's terms) to adapt to new demands in the environment. So, the results of changes in routines are evidenced in the dynamics of organizations (EED in our methodological approach).

One model that reframes the dynamic capabilities approach is that provided by Eisenhardt and Martin (2000), built on the resource-based view of the firm. The purpose of their model is to know how an organization's competitive edge can be maintained over time. The authors distinguish between two types of environments: moderately dynamic markets and high-velocity markets.[12] In this sense, moderately dynamic markets would be those distinctive for their stability, analytically detailed routines and predictable results, while high-velocity markets would be characterized by ambiguous structures, blurred boundaries where routines are linked to newly created knowledge and unpredictable results. It is the latter type of environment that we find most interesting from our approach. It is important to note the emphasis that

---

[12]The resource-based view of the firm "is enhanced by blending its usual path dependent strategic logic of leverage with a path-breaking strategic logic of change. [It] encounters a boundary condition in high velocity markets where the duration of competitiveness and advantage is inherently unpredictable, and dynamic capabilities are themselves unstable. Here the strategic imperative is not leverage but change" (Eisenhardt and Martin 2000: 1105).

Eisenhardt and Martin place on the velocity and/or rhythm of the changes. The dynamism of the environment is distinctive for the following properties (Davis et al. 2009):

- Velocity: The velocity at which new opportunities emerge (similar to epigenetic dynamics in our model).
- Complexity: The number of characteristics of an opportunity that must be correctly executed to better adapt to the environment
- Ambiguity: The degree of difficulty involved in distinguishing opportunities.
- Unpredictability: This would represent the amount of disorder in the flow of opportunities, which are less consistent with a previous framework.

Teece et al. (1997) introduce the concept of dynamic capabilities, defined as those which determine the firm's ability to integrate, build, and reconfigure internal and external resources/competences to address, and possibly shape, rapidly changing business environments. The analytical framework provided by the concept of EED seems to be a good fit with these authors' proposals. First, Teece (2007, 2010, 2012) recognizes changes in the initial routines, which in our EED model coincide with changes in the epigenome (i.e., initial routines of organizations). Second, Teece believes that changes in routines can and should be made in interaction with other external agents rather than exclusively as a result of an organization's own dynamic capabilities. Hence, Teece's dynamic capabilities acknowledge that the only way to adapt is through relationships with external agents from the environment (e.g., mergers and acquisitions, patent acquisitions, etc.). Finally, it is also important to underline the plasticity or adaptability property (Levinthal and Marino 2013), both in our model as well as in Teece's dynamic capabilities.

We find interesting that our methodological approach (i.e. EED) is close to what is defined as flexible and complex routines in the literature on strategic management. On the one hand, it strengthens our concept as it is similar in certain ways to said authors' development of dynamic routines. That is to say, it can be affirmed that a certain confluence is reached from different sources. Dynamic capabilities are initially related to EED insofar as there are routines that are inherited, replicated, and would fit with the most orthodox Darwinism although others would not. In other words, we can state that when studying genetics from the point of view of biology, there is a part of what would be the genome that is transmitted, although there are others that would form part of what we would call intense gene regulation which is envisaged in the EED methodology. Therefore, transmission is not so simple and mechanical, and elements that are not found in the body of instructions are sometimes transmitted. There was debate on this topic, above all in Europe. Hodgson (2010) and Pelikan (2010, 2012) took part and it was Pelikan's stricter approach which contended that, routines or instructions should be very clear from the point of view of genes, and even capable of being included in logical algorithms for it to be a valid analogy. Scholars such as Pelikan (2012) argue for the need to transpose all genetic instructions, mechanisms, and imprints into the biological analogy. Therefore, when neo-institutionalist actors talk about rules, for example, and others refer to routines, the analogy would not be valid if those rules or routines did not include a series of precise instructions.

As we discussed in the introduction to this chapter, three related points of attention could be cited as where the analyses concerning organizations' adaptation to changing environments should focus: the mechanics of change in routines; the necessary capabilities that organizations require; and the resulting dynamics observed in them. The EED approach also looks to analyze how the results of the dynamics followed by organizations impact the various systems (economic, social, etc.). This section will review the routines models that follow an evolutionary approach.

Table 3 provides an illustration of current authors' views on routines through the lens of the EED approach. Khalil (2012) points out a conceptual difference between instincts and routines, the first of which are abstract while routines are specific detailed remakes of abstract propositions written into instincts, which are practically unchanging. This dual concept addresses a biological interpretation made between genes and the environment (genotype/phenotype). Instincts have a very low degree of adaptive flexibility while this is relatively high for routines. In this analytical framework, routines would normally be in a state of adaptive variability and it could be inferred that "behavior ossifies in routines when the conditions in the environment continue to be stable" (Jablonka and Lamb 2005: 43).

Routines as adaptation to the environment will be of key interest in this study.[13] We also find the criteria used in Khalil's model interesting: economic rationality to assess adaptation of routines. The idea that agents are willing to reassess their routines on a permanent basis, which means that phenotype plasticity is not merely determined by genes, is also a relevant point in our opinion. However, we believe that plasticity must be very high to adapt to drastic changes in the environment.

If we take key developments in modern biology as a reference, it would then be possible to distinguish the coding genes which transmit (inherit) without changes from those which vary or are regulated. In this analytical framework, a conceptual separation between routines and the other hereditary and transmissible part without changes does not allow us to draw an analogy between instincts and genomes. Nor can instincts be called initial routines because they correspond solely to the part of the genome that is transmitted but not to the entire genome. In our view, the separation raised between instincts and routines reduces the operability and plausibility of Khalil's model. The difference between ostensive and performative routines (Pentland and Feldman 2005) may prove to be very useful.[14] At times, however, routines come from outside the organizations themselves and are put into practice in very short time periods. These "incorporated routines" may be very different from those observed in the organization itself and are carried out by different players. Agents' and organizations' actions and results do not always come from ostensive or performative routines; nor do they take shape as the same artifacts.

---

[13]We have seen that the epigenome is subject to the influence of the environment. Epigenetic inheritance is related to phenotype plasticity, which supports a Lamarckian interpretation (Jablonka and Lamb 2005).

[14]The concepts of ostensive and performative routines are defined in Table 3.

Table 3 Summary of views on routines seen from the EED perspective

Authors	Distinctive aspects of routines	Remarks from an EED perspective
Teece et al. (1997)	Introduce the concept of dynamic capabilities, defined as those which "determine the firm's ability to integrate, build and reconfigure internal and external resources/competences to address, and possibly shape, rapidly changing business environments"	The analytical framework provided by the concept of EED seems to be a good fit with these authors' proposals
Eisenhardt and Martin (2000)	Distinguish between: – moderately dynamic markets (characterized by stability, analytically detailed routines and predictable results) – high-velocity markets (characterized by ambiguous structures, blurred boundaries where routines are linked to newly created knowledge and unpredictable results)	They broaden the dynamic capabilities model High-velocity market environments are very similar to our understanding of changes in the environment
Zahra and George (2002)	Absorptive capacity defined as a set of organizational routines	The authors gave a positive response to our remarks on Massini et al. (2005)
Feldman and Pentland (2003)	They understand routines as permanent changes in systems The tasks to be carried out are analyzed with precision	The most relevant point seems to be the study of significant changes in organizations such as: their products, business scope and model, strategic decisions (including purchasing other firms' assets), etc.
Pentland and Feldman (2005)	They distinguish between: – Ostensive routines (abstract cognitive regularities to guide the actions of routines) – Performative routines (specific persons' actions at certain times – Artifacts (actions materialize as norms, written documents, procedures, algorithms, etc.)	Routines are sometimes brought in from external sources
Massini et al. (2005)	Two meanings: inherited genetic material and external routines They adapt to and learn from stakeholders and the external context	External purchase of knowledge assets and joint ventures should also be considered
Khalil (2012)	Routines classified as: – Instincts (written, unchanging instructions) – Routines (permanent remakes of instructions, plasticity)	The complexity of the genome is not taken into account
Levinthal and Marino (2013)	They introduce plasticity, which links changes to phenotypes The higher this is, the lower the replication capacity The concepts of mutation and phenotype are separated	Not all adaptation practices translate as improved evolution (result of the best practices)

Source Own elaboration

Another view complementary to Khalil's (2012) would be the one put forth by Massini et al. (2005) on the double perspective of routines: as inherited genetic material on the one hand and based on a permanent modification of routines over time to address unforeseen changes in the environment on the other. This adaptation rests mainly on external or meta routines in which each organization takes advantage of external experiences to increase its own level of cognition, learning from stakeholders (partners, suppliers, clients, etc.). In other words, external learning is combined with internal learning dynamics. From a Lamarckian view, the authors link internal absorptive capacity to investment in R&D and to the interrelation between external routines and the context of national innovation systems.

From our point of view, capturing external knowledge would often have to be rounded off with possible acquisition of assets found in other firms. This would be the case of assets linked to intellectual property and, at times, they might also be acquired by purchasing or entering into joint ventures with other firms. These strategic actions we have mentioned, and all external strategic moves, are related to internal learning. Thus, when agents act within their own organizations, they learn from the new concepts brought in from the outside in order to adapt to changes. Networking with technicians, engineers, and managers from other organizations that have been taken over or incorporated are important in these cases.[15]

Zahra and George (2002: 186) define absorptive capacity as "a set of organizational routines and processes by which firms acquire, assimilate, transform and exploit knowledge to produce a dynamic organizational capability." As Pentland et al. (2012: 1489) note, "if the routines display inertia, absorptive capacity will be low, learning will be low, and the capabilities of the organization may not be particularly dynamic." Pentland and Feldman (2008), acknowledge that routines are nonfixed generative systems. They are varied and undergo change on a permanent basis. Their model is functional and can be used for precise analysis of routines in actual practice.

However, an EED approach would not focus on precise knowledge of what each employee does, the workload or content as each task is carried out or the specific programs or formats formed by these tasks or results (artifacts). Our view centers on finding out the changes in organizations, in the products or services offered, in the field or market segments they enter, the range of necessary qualifications, the business scopes, the strategic assets needed and what makes them up (including those related to intellectual property) and the key strategic decisions made (including purchase/sale of assets and/or other firms).

Levinthal and Marino (2013) believe that selection is determined by adaptation, which is linked to learning and that the concept of routines emphasizes replication, as is clearly shown in Nelson and Winter's (1982) work. We find the importance placed on the concept of plasticity in the article by Levinthal et al. to be especially interesting: the higher the plasticity, the lower the replication capacity, which would weaken

---

[15]These perspectives have often been approached from the literature on open innovation (Chesbrough 2003; Huizingh 2011; Mortara and Minshall 2011; Van de Brande et al. 2009).

the stability property of the routines. On the contrary, without plasticity, organizations would not have the possibility to change and transform their resources to anticipate changes in circumstances (Teece et al. 1997). It can thus be said that the highest effectiveness in phenotype changes would be linked to plasticity. Levinthal and Marino (2013) propose some examples of adaptive internal selection such as policies to disseminate good practice and successful experiences which can be extended to other organizations.

In Levinthal et al.'s opinion, selection would, on the one hand, preserve the results of the best practices, while on the other, it would reject or suppress others which are found to be worse. From the viewpoint of EED, not all adaptive practices are due to evolutionary improvement. The results or consequences of epigenetic dynamics are more likely to be regressive and/or inefficient for the economic system in many cases. These dynamics can also clash with certain values or beliefs. Modulating beliefs and values would be analogous to silencing some genes, which would mean that a dynamic institutional/cultural framework in which certain values or beliefs disappear, giving way to new ones, could lead to the development of new epigenetic dynamics.

Lastly, as in the case of EED, Levinthal and Marino (2013) set up a clear methodological separation between mutation and phenotype. However, the latter is set within a slow adaptive process. Our model, nevertheless, places more emphasis on adaptive processes that may be fast and abrupt, but which would not be mutations in the strictest sense because epigenetic logic is very different from that of genetic mutation. In other words, these would not be failures in some gene that mutates, such as changes that occur in regulation of the genes themselves, and would therefore not be a strictly casual mutation.

Our approach identifies more closely with Teece et al.'s (1997) concept of dynamic capacities, which are defined as "higher level competences that determine the firm's ability to integrate, build, and reconfigure internal and external resources/competences to address, and possibly shape, rapidly changing business environments" (Teece 2007, 2010, 2012: 1). Teece distinguishes between what would actually be organizations' (group) and individuals' routinized behavior, and those dynamic capabilities that fall outside standardized analyses that search for the optimum situation. Teece identifies ordinary capabilities with routines that are a result of repetitive paths over time, which are embedded in organizations and employees and would be imprinted in the algorithms and heuristics of how businesses carry out and develop their everyday activities.

However, dynamic capabilities include changes and adaptation, often creating "fast moving competitive environments that require continuously modifying, and, if necessary, completely revamping what is doing so as to maintain a good fit with (and sometimes transform) the ecosystem that the enterprise occupies" (Teece 2012: 3). Business groups generate a specific framework of dynamic capabilities and competences which enables them to achieve a competitive edge over other groups in the face of rapid changes in the firms' environment. These may have to do with technology, access to markets, expectations and the competition's conditions, etc. These adaptations to the environment require renewing, rebuilding, and reconfiguring both the firm's internal and external competences. Management's coordination of firms' internal and external activities plays a key role. Learning processes are not exclusively generated internally but also occur in the framework of interorganizational relationships (Teece et al. 1997).

In Teece's words, "although some elements of dynamic capabilities may be embedded in the organization, the capability for evaluating and prescribing changes to asset configuration (both within and external to the organization) rests on the shoulders of top management" (Teece 2012: 4), which means that dynamic capabilities would preferably be found at high management levels. Top management is thus associated with the capacity to face the challenges stemming from changes or lack of adaptation to highly variable environments. When studying Internet industry groups, one of the most remarkable references besides their top management are the exceptional individuals that represent the firm's brand.

Although our approach is mainly limited to organizational routines, we do think our view could be broadened to consider what Pelikan (2010, 2012) put forth, which was that institutional norms, in North's terms (1990), may be equivalent to instructions. This new institutional orientation and North's ideas have a sociohistoric nature and are less appropriate for studying business organizations. From this perspective, analyses in terms of norms may be enriched when highlighting the change in institutional norms over relatively short time periods. It would make less sense to include the concept of learning on these coordinates of changes in norms whereas it is easier to see the concept of self-organization.

5 Introducing the EED Approach

In our opinion, epigenetic models may prove to be very useful when explaining the evolution of big Internet and telephone industry groups (Gómez-Uranga et al. 2013). However, the epigenetic analogy we use will not center on finding exhaustive precise parallels with the studies being carried out in biochemistry and genetics. We will focus solely on searching for relatively simple equivalencies that give a rough idea of the work being done on epigenetics in applied biology to ensure that we are using valid realistic analogies.

Epigenetics prompts the appearance of abnormalities or changes in what is programmed or initially encoded. In society, habits and routines are expressed in different ways, according to changes in external conditions. Change or modification of the aspects that accompany them may lead to changes in the normal expression of those business routines.

One of the properties that must be understood to conceptualize epigenetics is the uncertainty about the results of adaptation processes. When the analogy to "translate" the concept of epigenetics into the study of the dynamics in the Internet ecosystem is made, the main feature is the inability to make predictions about the potential dynamics that will characterize the evolution of the main Internet industry groups. That is why we can talk about highly uncertain environments.

The three-stage methodology proposed by the EED approach is as in Fig. 2.

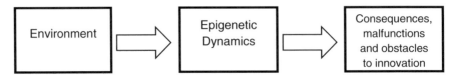

Fig. 2 The epigenetic economic dynamics (EED) approach. *Source* Own elaboration

5.1 Analysis of the Environment and Identification of the Genomic Instructions Which Are Transmitted Over Time

Methodologically, our point of departure is genetics. Thus, we look to identify the initial routines (i.e., the DNA) for each of the business groups included in our study. The first innovative products that are most closely identified with the business groups' initial activities would be those forming the essence of each company (i.e., most distinctive products) from its beginnings (e.g. Google's search engine, Microsoft's operating system, Apple's Mac, software, design and mobile phones, the possibility of downloading e-books in the case of Amazon and the social network concept developed by Facebook), and which made some of them market leaders. Their genome would also contain information about the routines and operating principles that would form their genetic footprint, such as: application of knowledge and technologies to enhance their value, market subordination, knowledge property management, separate assets (patents, brands, designs, copyrights, secrets, leadership), competition principles, profit goals, business models (e.g. free services, advertising, design), etc. These characteristics are assumed to be located in the DNA of these organisms or individual agents and are transmitted over time (similar to genes).

Business organizations' environments are exposed to great changes such as: developments in technologies[16]; fast-moving globalization, which implies considerable changes in business ecosystems (suppliers, customers, mergers and takeovers between groups, etc.) as well as power concentrated in the hands of big investors and higher competition between business organizations. Lastly, business models quickly become obsolete and product life cycles are increasingly shorter.

We will now focus on analyzing environmental influences such as: the evolution of the competition, evolution of technologies, changes in cultural patterns, etc. which affect each business group, institution, or agent and may lead to shifts in their initial routines. In this phase of epigenetic development, during which the relationship of the genotype is no longer a determining factor, conditions are created which may later lead to complications for the system and even extremely negative dynamics and dysfunctions in the companies themselves, affecting users and customers (Schubert 2012). As a result of the influence, introduction or addition

---

[16]Disruptive innovations place rapid limitations on what organizations are doing and how they usually carry out their activities (Gómez-Uranga et al. 2013).

of epigenetic factors, abnormalities and dysfunctions that stop innovation and/or block development of competition at different levels (intellectual property rights, abuse of monopoly power, etc.) may arise.

These business groups face an environment with the following characteristics: intense increase in intergroup competition, exponential growth of the markets and users in other (related) business areas, a high demand for innovation, increase in the number of applications and their content, fast multivectorial technological change and rapid planned obsolescence (Miao 2011), modularity in the behavior of business ecosystems, higher advertising and marketing expenses, and an exponential increase in the patent portfolio. The rapidly growing number of users puts such pressure on the demand that it leads to major changes in the dynamics of the environment, prompting growth and thousands of new user applications. Above all, the entire process is occurring extremely fast. The previous actions are due to the high variability of these big groups' environments. The drastic rapid changes in the environments and the high-velocity markets in which they operate, such as those that come from shifts in science and technology paradigms, have a considerable impact on the dynamics of these business groups (Eisenhardt and Martin 2000; Wirtz et al. 2007). These environments are undergoing extraordinarily fast changes in the fields of technologies, business logic, intensified intergroup competition, ways to access knowledge on a patents scene characterized by saturation and litigation, and above all, dramatic growth in user demands for existing products and services on different platforms.[17]

5.2 Identification of Epigenetic Dynamics

This second stage focuses on analyzing the changes observed in the business ecosystem in response to influences from the environment. Some examples would be evolution of the competition and technologies, changes in cultural patterns, etc. that influence each business group, institution, or agent and can induce change, variation, or add functions to their corresponding DNAs.

As mentioned before, business groups are conditioned by their environment, as a result of which "genetic disorders" may be created. These changes resulting from the environment where the business groups operate build new paths that become part of their new identity. Previous identities are modified as these new ones are transmitted or replicated over time (David 1985, 1994). However, these changes occur abruptly rather than gradually, as if they were mutations. It is worth noting that these mutations (like learning) are deliberate. In other words, they do not happen by chance as we might deduce from the Darwinism that has dominated evolutionary thought to the present time.

---

[17]We consider that alternative approaches such as that introduced by Geels (2002) on technological and sustainability transitions could also be compatible with the EED, in particular when addressing the analysis of the changes in the environment.

What we find most enlightening in our analysis is epigenetic dynamics as a response to changes in the environment. Put differently, how organizations adapt and what dynamics they adapt to. At this point, we find this adaptation has some core factors, for instance, these business groups' purchases and acquisitions (see Chapter "Epigenetic Economics Dynamics in the Internet Ecosystem"). Some of these groups' frenetic dynamics are revolutionizing the entire Internet industry. In this section, we offer some examples and, in a certain manner, highlight the big differences that exist when establishing more or less acceptable competition because the power of these groups is somehow so dominant (economically and financially) that it practically wipes competitors (and even potential competitors) off the map.

Epigenetic dynamics follow an economic rationality, which means that these groups need (as a result of the changes which have occurred in their environments) to sustain profit growth. Among other reasons, this is to meet their investors' demands for profitability and justify their investments. At the same time, these groups need to obtain significant results in innovative terms. Improved innovation makes them more successful when competing in these disruptive environments. Schumpeter's dialectics of entrepreneur/innovator are perfectly applicable and hence give meaning to these epigenetic dynamics.

In order to adapt to this environment, business groups sometimes have to acquire external knowledge since they cannot find it in-house (Mortara and Minshall 2011). This external knowledge very often needs to be supplemented with the acquisition of other firms' assets (i.e. patents, acquisition of companies, joint venture agreements, etc.). As a matter of fact, the acquisitions made by Facebook between 2005 and 2014 totalled more than 23 billion USD, some of them being particularly noteworthy, like Instagram (1 billion USD), Whatsapp (19 billion USD), or Oculus (2 billion). In the case of Google, the number of acquisitions between 2003 and 2014 rose to 153, representing a total investment of 137,000,000 billion USD. Therefore, the financial surplus of business groups is essential to acquire knowledge which is not available internally and allow them to adapt to the environment and compete in it.

The epigenetic framework which is gradually designed for each organization or agent also affects how it works and the result of its main function (i.e. DNA). However, veering from the path marked by the DNA is not so simple, and carries a price. At the time, Microsoft did not consider it a good business move to penetrate the search engine segment so its first efforts on the Internet centered on browsers, firstly competing against Netscape and later against Mozilla (Cleland and Brodsky 2011; Suárez Sánchez-Ocaña 2012). Later, in a 'natural evolution' framework, Google absorbed other thematic search engine companies such as Aardvark, Metawen, Plinkart, ITA, Like.com, etc. This prompted new dynamics and evolution in these agents, leaving new fingerprints that are transmitted over time. These epigenetic dynamics are perceived as an institution which lasts for a long period of time. A case in point is Google. The group works in many different fields other than search engines, but will always be identified with that main function, which is in its DNA. However, the means through which they are passed on are not so easy to identify as DNA.

Iansiti and Richards (2006) draw an analogy between competition and evolution of the species in the sense that some animal species 'run a race' to adapt in their evolution (Dawkins 1976, 1982, 1983). This enables them to defend themselves from their predators to avoid their extinction as a species. From an evolutionary perspective, for the large Internet industry groups competition means permanently resizing and readapting to maintain an identity, a place on the market which may sometimes be the leading position, and which requires strategies to take over and merge with other groups. We could say that these business groups' genome contains the need to compete in order to maintain their leadership, but also to survive (as a group).

Some of the characteristics of these epigenetic dynamics include: massive acquisition of small firms and/or their intellectual property (i.e., patent portfolio) to block potential structural changes and to defend themselves from competition; aggressive acquisition strategies to sustain profit growth, presence on global markets and gain access to new technologies and innovations; asymmetric negotiations between large business groups, application developers and content providers; entry of large business groups in activities not related to their original purpose (DNA); high-entry barriers posed by large incumbents; and financial strength as the main protective industrial instrument.

5.3 Consequences (in Terms of Innovation) as a Result of Epigenetic Factors

The third stage leads to conclusions about the abnormalities, malfunctions, or obstacles to innovation, and/or blockage of the competition's development at certain levels (intellectual property rights, abuse of monopoly power, etc.) that are observed in the ecosystem and which may arise as a result of the influence, introduction, or addition of epigenetic factors. Some of the implications or consequences of the previous epigenetic dynamics include: existence of a gap between R&D investments and patenting results; distorted patenting rationale; excessive transaction (and litigation) costs; high-entry barriers to SME patenting; problems in standards definition and development; overload in patent offices and regulating agencies due to the existing patenting inflation.

Patents are one of the strongest environmental properties of the Internet ecosystem. The field of patents shows just how fierce the competition is. Lawsuits for patent infringement or violation are quite common (Cunningham 2011). Companies sometimes seriously alter competition through their lawsuits, filing claims to stop the sale of their rivals' products. In theory, patents ensure progress and technology advances. In practice, they have become a battlefield for cross-claims which questions one of the key objectives of patents systems. Patents are now being used to hinder competitors' growth (The Antitrust Bulletin 2005). The meaning of patents has changed: they used to be the result of innovation and companies could pay for the use of license rights, but now they seek exclusive rights so as to include them in their ecosystems and thus hinder rivals companies' growth (i.e., blocking the potential innovation capacity of competitors rather than creating the necessary incentives to innovate).

For the main Internet business groups, patents are a source of big expense, especially as regards human resources. Keeping up a patent and license portfolio through litigation involves huge expenses. Armies of engineers and lawyers spend more time working on patents than on what is strictly R&D. Furthermore, in an Internet economy, increasingly larger proportions of revenues must be devoted to R&D to confront stiff competition. In addition to these huge expenses, litigation acts as a disincentive for innovators.

It is also important to take into account the impact caused by inefficiencies in the patent system as per the high price of the end product/service as well as higher transaction costs resulting from patenting expenses and related lawsuits (Encaoua and Madiès 2012). This inefficiency implies that products/services take longer to reach the market because of the time involved in patenting and the lawsuits which may result. Bessen and Meurer (2008) state that the intellectual property rights system has failed as a form of protection and information for companies in the USA. Lawsuits for infringement of intellectual property could even be affecting the share price of different business groups. Although the situation of patents and incentives for innovation varies according to the industry, software patents are very abstract and poorly defined. This makes it much more complicated to achieve reasonably efficient market contracts (Bessen and Meurer 2008, 2012). Therefore, we could say that market failure is due to poorly defined property rights. All of these issues lead us to ask if patents systems can no longer fulfill their primary objectives.

Efficient patent policy enables companies to compete in better conditions. Thus, the need for antitrust and competition oversight bodies, or the Department of Justice in the USA and the European Commissioner for Competition to act. The Department of Justice itself brought out a guide focused on a flexible approach to the most common problems in June 2011 (Department of Justice 2012). Possible remedies include mandatory licensing on fair and legal terms, acting to stop retaliation from merged firms and also banning certain contracting practices (Fischer and Henkel 2012; Knable Gotts and Sher 2012; Turner 2011).

Sharp growth in the number of patent applications, as well as their voluminosity (size and scope) has been especially noticeable in patent offices in recent years (van Zeebroeck et al. 2009; Gómez-Uranga et al. 2014). Similarly, the acquisitions of large patent portfolios from smaller companies (i.e. start-ups) have also given more market power to the largest Internet business groups. These acquisitions sometimes consist of thousands of patents, giving the buyers leverage to block the growth of these new entrants.

The EED approach therefore allows the inclusion of the consequences of epigenetic dynamics, which so far have been addressed in a limited manner in the case of patents (Gómez-Uranga et al. 2014). However, we still need to expand the analysis of the consequences to other areas that we believe may be as interesting or more that the patent system.

Table 4 includes a diagram of the epigenetic process. We believe this model will enable us to establish the adaptations/relationships between the environment and the dynamics (which we call epigenetic dynamics) and also the results of these (through consequentialist logic) which may lead to improvements in the

Table 4 An epigenetic approach to understanding the economic and social impact of ecosystem dynamics

A. The ENVIRONMENT	
Intense increase in intergroup competition	
Exponential growth of the markets and users in other (related) business areas	
Increase in the number of applications and their content	
Fast multivectorial technological change and planned obsolescence	
Modularity in the behavior of business ecosystems	
Exponential increase in advertising as a share of turnover	
Increase in marketing expenses	
Exponential increase in patent portfolios	
Industry, market, and institutional structures	
B. EPIGENETIC dynamics in response to the ENVIRONMENT	
High entry barriers posed by large incumbents in certain industry niches	
Risk-averse industrial strategies implemented by large companies	
Massive acquisition of patent portfolios	
Financial strength as the main defensive industrial instrument	
Asymmetric negotiations between large business groups, application developers, and content providers	
Acquisition of small firms and/or their intellectual property to block potential structural changes and to defend themselves from competition	
Aggressive acquisition strategies to sustain profit growth and their presence on global markets and to gain access to new technologies and innovations	
Entry of large business groups in activities not related to their original purpose (DNA)	Some examples: OS (Android), IOS (Apple), Symbian (Nokia), Bada (Samsung), new tablets (Nexus 7-Google, iPad mini-Apple, Kindle Fire HD-Amazon, Galaxy tab. 27.0-Samsung)
C. An illustration of the consequences, malfunctions and implications of epigenetic dynamics	
Conservative and defensive innovation strategies of large corporations	
Blocking competition—difficulties for new firms' to grow—risk of lock-in	
R&D gap/patenting results	
Distorted patenting rationale	
Excessive transaction (and litigation) costs	
High entry barriers to SME patenting	
Problems in standards definition and development	
Overload in patent offices and regulating agencies	
Efficiency of business firms	Knowledge and technology transfer
Economic growth	Growth of inequalities
Innovation in organizations	Employment generation
Innovation in territorial spaces	Property right regulations (particularly intellectual)
Innovative capacity of small firms, start-ups or individuals	Sectoral and global economic competition
Innovation-friendly environment	Inequalities in access to information and knowledge
Relative negotiation power of users and customers unveiled	Monopolistic entry barriers
Fiscal justice	Tax evasion
Moral values exposed	etc.

Source Own elaboration based on Gómez-Uranga et al. (2014)

system, or in this case, functional failures, system failures, and in other cases, moral problems. In Table 4, we place strong emphasis on the implications of the epigenetic dynamics derived from the need to adapt to the environment.[18] In this sense, we consider that epigenetic dynamics may result in the improvement or deterioration (i.e., degradation) of the following objectives.

As was pointed out previously, the environment is one of the main factors influencing the behavior of business groups and territories. The environment conditions business groups and regions, and may even damage their very origin (i.e., their genotype, their initial routines), for instance: their image, reputation, action areas, etc.; values that nurture the foundations of their acceptance (as well as that of their products/services), their growth, etc. These changes stemming from the environment in which business groups operate lead to the formation of new paths to follow.

References

Abatecola, G., Belussi, F., Breslin, D., & Filatotchev, I. (2015). Darwinism, organizational evolution and survival: Key challenges for future research. *Journal of Management & Governance*, 1–17.

Aldrich, H. E., Hodgson, G. M., Hull, D. L., Knudsen, T., Mokyr, J., & Vanberg, V. J. (2008). In defence of generalized Darwinism. *Journal of Evolutionary Economics, 18*, 577–596.

Ansoff, H. I., & Sullivan, P. A. (1993). Optimizing profitability in turbulent environments: A formula for strategic success. *Long Range Planning, 26*(5), 11–23.

Argote, L. (1999). *Organizational learning: Creating, retaining, and transferring knowledge.* New York: Kluwer Academic Publishers.

Arthur, W. B. (1999). Complexity and the economy. *Science, 284*, 107–109.

Ayala, F. J. (2013). Aquí se producen científicos y el resto del mundo se beneficia de ellos. *Revista Kampusa, 73*, 10–11.

Basole, R. C. (2009). Structural analysis and visualization of ecosystems: A study of mobile device platforms. In *Proceedings of the Fifteenth Americas Conference on Information Systems*, San Francisco, California, August 6–9, 2009.

Bessen, J., & Meurer, M. J. (2008). *Patent failure: How judges, bureaucrats, and lawyers put innovators at risk.* Woodstock: Princeton University Press.

Bessen, J., & Meurer, M. J. (2012). The direct costs from NPE disputes. Working paper 12–34. Boston University School of Law.

Bingham, C. B., & Eisenhardt, K. M. (2011). Rational heuristics: The 'simple rules' that strategists learn from process experience. *Strategic Management Journal, 32*, 1437–1464.

Bingham, C. B., & Eisenhardt, K. M. (2014). Heuristics in strategy and organizations: Response to Vuori and Vuori. *Strategic Management Journal, 35*(11), 1698–1702.

Boschma, R. (2015). Towards an evolutionary perspective on regional resilience. *Regional Studies, 49*(5), 733–751.

Boschma, R. A., & Frenken, K. (2006) Why is economic geography not an evolutionary science? Towards an evolutionary economic geography. *Journal of Economic Geography, 6*(3), 273–302.

Boschma, R. A., & Frenken, K. (2009). The spatial evolution of innovation networks. A proximity perspective. In R. Boschma, & R. Martin (Eds.), *The handbook of evolutionary economic geography* (pp. 120–135). Cheltenham: Edward Elgar.

---

[18]For an illustration of the main features characterizing items A (environment) and B (epigenetic dynamics) see Gómez-Uranga et al. (2013).

Boschma, R. A., & Martin, R. (2007). Constructing an evolutionary economic geography. *Journal of Economic Geography, 7*, 537–548.

Boschma, R., & Martin, R. (2010). The aims and scope of evolutionary economic geography. In R. Boschma & R. Martin (Eds.), *The handbook of evolutionary economic geography* (pp. 3–39). Cheltenham: Edward Elgar.

Breslin, D. (2011). Reviewing a generalized Darwinist approach to studying socio-economic change. *International Journal of Management Review, 13*, 218–235.

Campbell, D. T. (1965). Variation and selective retention in socio-cultural evolution. In H. R. Barringer, G. I. Blankstein, & R. W. Mack (Eds.), *Social change in developing areas: A reinterpretation of evolutionary theory* (pp. 19–49). Cambridge, MA: Schenkman.

Canetti, E. (2003). Epigenética: una explicación de las enfermedades hereditarias. *Perinatología y Reproducción humana, 17*, 57–60.

Carey, N. (2012). *The epigenetics revolution. How modern biology is rewriting our understanding of genetics, disease and inheritance.* New York: Columbia University Press.

Carroll, S. B. (2005). *Endless forms most beautiful: The new science of evo devo and the making of the animal kingdom.* New York: W.W. Norton.

Cavagnari, B. M. (2012). Regulación de la expresión génica: cómo operan los mecanismos epigenéticos. *Archivos argentinos de pediatría, 110*(2), 132–136.

Chesbrough, H. (2003). *Open innovation. The new imperative for creating and profiting from technology.* Boston, MA: Harvard Business School Publishing Corporation.

Cleland, S., & Brodsky, I. (2011). *Search and destroy: Why you can't trust Google Inc.* St. Louise, Missouri: Telescope books.

Cooke, P., Gomez Uranga, M., Etxebarria, G. (1997). Regional innovation systems: Institutional and organizational dimensions. *Research Policy, 26*, 475–491.

Cordes, C. (2006). Darwinism in economics: From analogy to continuity. *Journal of Evolutionary Economics, 16*(5), 529–541.

Corkery, T., Forder, J., Svantesson, D., & Mercuri, E. (2015). Taxes, the internet and the digital economy. *Revenue Law Journal, 23*(1), Article 7.

Corning, P. A. (2002). The re-emergence of "emergence": A venerable concept in search of a theory. *Complexity, 7*(6), 18–30.

Cunningham, S. (2011). Mobile patent suits: Graphic of the day. http://blog.thomsonreuters.com/index.php/mobile-patent-suits-graphic-of-the-day. Accessed 3 October 2015.

Darwin, C. R. (1859). *On the origin of species by means of natural selection, or the preservation of favored races in the struggle for life.* London: Murray.

Darwin, C. R. (1871). *The descent of man, and selection in relation to sex.* London: Murray.

David, P. A. (1985). Clio and the economics of QWERTY. *American Economic Review, 75*(2), 332–337.

David, P. A. (1994). Why are institutions the 'carriers of history? Path dependence and the evolution of conventions, organizations and institutions. *Structural Change and Economic Dynamics, 5*(2), 205–220.

Davies, P. (2002). La flecha del tiempo. *Investigación y Ciencia*, November 8–13.

Davis, J. P., Eisenhardt, K. M., & Bingham, C. B. (2009). Optimal structure, market dynamism, and the strategy of simple rules. *Administrative Science Quarterly, 54*, 413–452.

Dawkins, R. (1976). *The selfish gene.* Oxford: Oxford University Press.

Dawkins, R. (1982). *The extended phenotype: The long reach of the gene.* Oxford: W.H. Freeman and Company.

Dawkins, R. (1983). Universal Darwinism. In D. S. Bendall (Ed.), *Evolution from molecules to man* (pp. 403–425). Cambridge: Cambridge University Press.

Deighton, J. A., & Kornfeld, L. (2013). Amazon, Apple, Facebook and Google. *Harvard Business School Case,* 513-060.

Department of Justice. (2012). Statement of the Department of Justice's Antitrust Division on its decision to close its investigations of Google Inc.'s Acquisition of Motorola Mobility Holdings Inc. and the Acquisitions of certain patents by Apple Inc., Microsoft Corp. and Research in Motion Ltd. 13th February 2012. http://www.justice.gov/opa/pr/2012/February/12-at-210.html. Accessed 3 October 2015.

Dosi, G. (1982). Technological paradigms and technological trajectories: A suggested interpretation of the determinants and directions of technical change. *Research Policy, 11*(3), 147–162.

Eisenhardt, K. M., & Martin, J. A. (2000). Dynamic capabilities: What are they? *Strategic Management Journal, 21*(10–11), 1105–1121.

Encaoua, D., & Madiès, T. (2012). Dysfunctions of the patent system and their effects on competition. http://hal.archives-ouvertes.fr/docs/00/74/07/16/PDF/ENCAOUA-MADIES_DYSFUNCTIONS_.pdf. Accessed 3 October 2015.

Essletzbichler, J., & Rigby, D. L. (2010). Generalized Darwinism and evolutionary economic geography. In R. Boschma & R. Martin (Eds.), *The handbook of evolutionary economic geography* (pp. 43–61). Cheltenham: Edward Elgar.

Etkin, J., & Schvarstein, L. (1995). *Identidad de las Organizaciones. Invariancia y Cambio.* Buenos Aires: Paidós.

Feldman, M. S., & Pentland, B. T. (2003). Reconceptualizing organizational routines as a source of flexibility and change. *Administrative Science Quarterly, 48*, 94–118.

Fischer, T., & Henkel, J. (2012). Patent trolls on markets for technology: An empirical analysis of NPEs' patent acquisitions. *Research Policy, 41*, 1519–1533.

Folke, C., Carpenter, S. R., Walker, B., Scheffer, M., Chapin, T., & Rockström, J. (2010). Resilience thinking: Integrating resilience, adaptability and transformability. *Ecology and Society*, 15(4), article 20.

Francis, R. C. (2011). *Epigenetics. How the environment shapes our genes.* New York: W.W. Norton & Company.

Fransman, M. (2014). *Models of innovation in global ICT firms: The emerging global innovation ecosystems.* Luxembourg: European Union.

Freeman, C. (2002). Continental, national and sub-national innovation systems—complementarity and economic growth. *Research Policy, 31*(2), 191–211.

Frenken, K., & Boschma, R. A. (2007). A theoretical framework for evolutionary economic geography: Industrial dynamics and urban growth as a branching process. *Journal of Economic Geography, 7*(5), 635–649.

García Azkonobieta, T. (2005). Evolución, desarrollo y (auto)organización: Un estudio sobre los principios filosóficos de la evo-devo. PhD thesis. San Sebastian: University of the Basque Country.

Geels, F. (2002). Technological transitions as evolutionary reconfiguration processes: A multi-level perspective and a case-study. *Research Policy, 31*, 1257–1274.

Geels, F. (2014). Reconceptualising the co-evolution of firms-in-industries and their environments: Developing an inter-disciplinary triple embeddedness framework. *Research Policy, 43*(2), 261–277.

Gilbert, S. F. (2001). Ecological developmental biology: Developmental biology meets the real world. *Developmental Biology, 233*, 1–22.

Gilbert, S. F., & Epel, D. (2009). *Ecological developmental biology.* Sunderland, MA: Sinauer Associates.

Gillham, N. W. (2001). Evolution by jumps: Francis Galton and William Bateson and the mechanism of evolutionary change. *Genetics, 159*, 1383–1392.

Gleick, J. (1987). *Chaos: Making a new science.* New York: Penguin books.

Gómez-Uranga, M., Zabala-Iturriagagoitia, J. M., & Miguel, J.C. (2013). Evolutionary epigenetic economics: How to better understand the trends of big internet groups. SSRN. http://papers.ssrn.com/sol3/papers.cfm?abstract_id=2200421. Accessed 3 October 2015.

Gómez-Uranga, M., Miguel, J. C., & Zabala-Iturriagagoitia, J. M. (2014). Epigenetic economic dynamics: The evolution of big Internet business ecosystems, evidence for patents. *Technovation, 34*(3), 177–189.

Gottlieb, G. (2001). A developmental psychobiological systems view: Early formulation and current status. In S. Oyama, P. E. Griffiths, & R. D. Gray (Eds.), *Cycles of contingency: Developmental systems and evolution.* Cambridge, MA: MIT Press.

Griffiths, P. E. (1996). Darwinism, process structuralism and natural kinds. *Philosophy of Science, 63*, S1–S9.

Griffiths, P. E., & Gray, R. D. (2005). Discussion: Three ways to misunderstand developmental systems theory. *Biology and Philosophy, 20*, 417–425.

Heckemeyer, J. H., Richter, K., Spengel, C. (2014). Tax planning of R&D intensive multinationals. ZEW Discussion Papers, 14-114.

Hernández-Martínez, A. G. (2006). La decisión y su relación con el tiempo: estrategia, procesos e identidad. *Revista Facultad de Ciencias Económicas: Investigación y Reflexión, 14*(1), 23–43.

Herstatt, C., & Kalogerakis, K. (2005). How to use analogies for breakthrough innovations. *International Journal of Innovation and Technology Management, 2*(3), 331–347.

Hodgson, G. M. (1993). *Economics and evolution: Bringing life back into economics.* Cambridge: Cambridge Polity Press.

Hodgson, G. M. (2009). Agency, institutions and Darwinism in evolutionary economic geography. *Economic Geography, 85*(2), 167–173.

Hodgson, G. M. (2010). Darwinian coevolution of organizations and the environment. *Ecological Economics, 69*(4), 700–706.

Hodgson, G. M. (2012). The mirage of microfoundations. *Journal of Management Studies, 49*, 1389–1395.

Hodgson, G. M., & Knudsen, T. (2004). The firm as an interactor: Firms as vehicles for habits and routines. *Journal of Evolutionary Economics, 14*(3), 281–307.

Hodgson, G. M., & Knudsen, T. (2006a). Why we need a generalized Darwinism, and why a generalized Darwinism is not enough. *Journal of Economic Behavior and Organization, 61*, 1–19.

Hodgson, G. M., & Knudsen, T. (2006b). Dismantling Lamarckism: Why descriptions of socio-economic evolution as Lamarckian are misleading. *Journal of Evolutionary Economics, 16*, 343–366.

Hodgson, G. M., & Knudsen, T. (2012). Agreeing on generalised Darwinism: A response to Pavel Pelikan. *Journal of Evolutionary Economics, 22*(1), 9–18.

Holland, J. H. (1998). *Emergence: From chaos to order.* Redwood City, CA: Addison-Wesley.

Huizingh, E. K. R. E. (2011). Open innovation: Sate of the art and future perspectives. *Technovation, 31*, 2–9.

Iansiti, M., & Richards, G. L. (2006). The information technology ecosystem: Structure, health, and performance. *The Antitrust Bulletin, 51*(1), 77–110.

Jablonka, E., & Lamb, M. J. (2005). *Evolution in four dimensions: Genetic, epigenetic, and symbolic variation in the history of life.* Cambridge, MA: MIT Press Books.

Jing, Z., & Xiong-Jian, L. (2011). Business ecosystem strategies of mobile network operators in the 3G era: The case of China Mobile. *Telecommunications Policy, 35*, 156–171.

Kauffman, S. (1993). *Origins of order: Self-organization and selection in evolution.* Oxford: Oxford University Press.

Kauffman, J. B., Case, R. L., Lytjen, D., Otting, L., & Cummings, D. L. (1995). Ecological approaches to riparian restoration in northeastern Oregon. *Restoration and Management Notes, 13*, 12–15.

Khalil, E. L. (2012). Are instincts hardened routines? A radical proposal. Monash University. Department of Economics, Discussion Paper 25/12. Retrieved October 3, 2015 from http://www.buseco.monash.edu.au/eco/research/papers/2012/2515areinstinctskhalil.pdf.

Kitano, H. (2004). Biological robustness. *Nature Reviews Genetics, 5*, 826–837.

Knable Gotts, I., & Sher, S. (2012). The particular antitrust concerns with patent acquisitions. *Competition Law International*, August, 30–38.

Lara, B. (1991). *La decisión, un problema contemporáneo.* Madrid: Editorial Espasa Calpe S.A.

Levinthal, D. A. (1997). Adaptation on rugged landscapes. *Management Science, 43*(7), 934–950.

Levinthal, D. A., & Marino, A. (2013). *Three facets of organisational adaptation: Selection, variety, and plasticity.* The Wharton School: University of Pennsylvania.

Levit, G. S., Hossfeld, U., & Witt, U. (2011). Can Darwinism be 'generalized' and of what use would this be? *Journal of Evolutionary Economics, 21*, 545–562.

Levitt, B., & March, J. (1988). Organizational learning. *Annual Review of Sociology, 14*(1), 319–340.

Lewontin, R. C. (1982). Organism and environment. In H. Plotkin (Ed.), *Learning, development, culture* (pp. 151–170). New York: Wiley.

Lewontin, R. C. (1983). Gene, organism and environment. In D. S. Bendall (Ed.), *Evolution: From molecules to men* (pp. 273–285). Cambridge: Cambridge University Press.

Li, J. (2014). Protecting the tax base in the digital economy. Papers on selected topics in protecting the tax base of developing countries. Paper No. 9, June 2014. New York: United Nations.

Lundvall, B. -Å. (1992). *National innovation systems: Towards a theory of innovation and interactive learning*. London: Pinter.

Martin, R. (2012a). Regional economic resilience, hysteresis and recessionary shocks. *Journal of Economic Geography, 12*, 1–32.

Martin, R. (2012b). (Re)Placing path dependence: A response to the debate. *International Journal of Urban and Regional Research, 26*(1), 179–192.

Martin, R., & Sunley, P. (2007). Complexity thinking and evolutionary economic geography. *Journal of Economic Geography, 7*, 573–601.

Martin, R., & Sunley, P. (2014). Towards a developmental turn in evolutionary economic geography? *Regional Studies, 49*(5), 712–732.

Massini, S., Lewin, A. Y., & Greve, H. R. (2005). Innovators and imitators: Organizational reference groups and adoption of organizational routines. *Research Policy, 34*(10), 1550–1569.

Masuelli, R. W., & Marfil, C. F. (2011). Variabilidad epigenética en plantas y evolución. *Journal of Basic and Applied Genetics, 22*(1), 1–8.

Mayr, E. (1988). *Toward a new philosophy of biology: Observations of an evolutionist*. Cambridge, MA: Harvard University Press.

Mayr, E. (1991). *One long argument: Charles Darwin and the genesis of modern evolutionary thought*. Cambridge, MA: Harvard University Press.

Miao, C.-H. (2011). Planned obsolescence and monopoly undersupply. *Information, Economics and Policy, 23*(1), 51–58.

Mitleton-Kelly, E. (2003). Ten principles of complexity & enabling infrastructures. In E. Mitleton-Kelly (Ed.), *Complex systems and evolutionary perspectives of organisations: The application of complexity theory to organizations* (pp. 23–50). Elsevier.

Mortara, L., & Minshall, T. (2011). How do large multinational companies implement open innovation? *Technovation, 31*(10–11), 586–597.

Nelson, R. R. (1995). Recent evolutionary theorizing about economic change. *Journal of Economic Literature, 33*(1), 48–90.

Nelson, R. R. (2005). Evolutionary social science and universal darwinism. CCS Working Paper 5. Center on Capitalism and Society, The Earth Institute at Columbia University Working Papers Series.

Nelson, R. R. (2007). Universal Darwinism and evolutionary social science. *Biology and Philosophy, 22*(1), 73–94.

Nelson, R. R., & Winter, S. G. (1982). *An evolutionary theory of economic change*. Cambridge, MA: Harvard University Press.

North, D. C. (1990). *Institutions, institutional change and economic performance*. Cambridge: Cambridge University Press.

Oyama, S., Griffiths, P. E., & Gray, R. D. (2001). *Cycles of contingency: Developmental systems and evolution*. Cambridge, MA: MIT Press.

Pelikan, P. (2010). Evolutionary developmental economics: How to generalize Darwinism fruitfully to help comprehend economic change. *Journal of Evolutionary Economics, 21*, 341–366.

Pelikan, P. (2012). Agreeing on generalized Darwinism: A response to Geoffrey Hodgson and Thorbjörn Knudsen. *Journal of evolutionary economics, 22*, 1–8.

Pendall, R., Foster, K. A., Cowell, M. (2010). Resilience and regions: Building understanding of the metaphor. *Cambridge Journal of Regions, Economy and Society, 3*, 71–84.

Pentland, B. T., & Feldman, M. S. (2005). Organizational routines as a unit of analysis. *Industrial and Corporate Change, 14*(5), 793–815.

Pentland, B. T., & Feldman, M. S. (2008). Issues in empirical field studies of organizational routines. In M. C. Becker (Ed.), *Handbook of organizational routines* (pp. 281–300). Cheltenham: Edward Elgar.

Pentland, B. T., Feldman, M. S., Becker, M. C., & Liu, P. (2012). Dynamics of organizational routines: A generative model. *Journal of Management Studies, 49*(8), 1484–1508.

Petrosky, T. Y., & Prigogine, I. (1990). Laws and events: The dynamical basis of self-organization. *Canadian Journal of Physics, 68*(9), 670–682.

Pike, A. (2002). Task forces and the organisation of economic development: The case of the North East region of England. *Environment and Planning C, 20*, 717–739.

Pike, A., Dawley, S., & Tomaney, J. (2010). Resilience, adaptation and adaptability. *Cambridge Journal of Regions, Economy and Society, 3*, 59–70.

Pohl, J. (1999). Some notions of complex adaptive systems and their relationship to our world. Presented at InterSymp-99, Advances in Collaborative Decision-Support Systems for Design, Planning and Execution (pp. 9–24). Baden-Baden, Germany, August 2–7, 1999.

Robert, J. S., Hall, B. K., & Olson, W. M. (2001). Bridging the gap between developmental systems theory and evolutionary developmental biology. *BioEssays, 23*, 954–962.

Schubert, C. (2012). Is novelty always a good thing? Towards an evolutionary welfare economics. *Journal of Evolutionary Economics, 22*(3), 585–619.

Stoelhorst, J.W. (2008). The explanatory logic and ontological commitments of generalized darwinism. *Journal of Economic Methodology, 15*(4), 343–363.

Suárez, D. (2014). Persistence of innovation in unstable environments: Continuity and change in the firms' innovative behavior. *Research Policy, 43*(4), 726–736.

Suárez Sánchez-Ocaña, A. (2012). *Desnudando a Google: La inquietante realidad que no quieren que conozcas*. Barcelona: Editorial Deusto.

Teece, D. J. (2007). Explicating dynamic capabilities: The nature and microfoundations of (sustainable) enterprise performance. *Strategic Management Journal, 28*, 1319–1350.

Teece, D. J. (2010). Technological Innovation and the theory of the firm: The role of enterprise-level knowledge, complementarities and (dynamic) capabilities. In B. H. Hall & N. Rosenberg (Eds.), *Handbook on the economics of innovation* (pp. 679–730). North-Holland: Elsevier.

Teece, D. J. (2012). Dynamic capabilities: Routines versus entrepreneurial action. *Journal of Management Studies, 49*(8), 1395–1401.

Teece, D. J., Pisano, G., & Shuen, A. (1997). Dynamic capabilities and strategic management. *Strategic Management Journal, 18*(7), 509–533.

Turner, J. L. (2011). *Patent thickets, trolls and unproductive entrepreneurship*. Athens, GA: Department of Economics, Terry College of Business, University of Georgia.

Van de Brande, V., De Jong, J. P. G., Vanhaverbeke, W., & De Rochemont, M. (2009). Open innovation in SMEs: Trends, motives and management challenges. *Technovation, 29*, 423–437.

Van Gundy, A. B. (1981). *Techniques of structured problem solving*. New York: Van Nostrand Reinhold Company.

Van Zeebroeck, N., Pottelsberghe, Van, de la Potterie, B., & Guellec, D. (2009). Claiming more: The increased voluminosity of patent applications and its determinants. *Research Policy, 38*(6), 1006–1020.

Vosniadou, S., & Ortony, A. (1989). *Similarity and analogical reasoning*. Cambridge: Cambridge University Press.

Waddington, C. H. (1947). *Organisers and genes*. Cambridge, MA: Cambridge University Press.

Waddington, C. H. (1953). Genetic assimilation of an acquired character. *Evolution, 7*, 118–126.

Weitzman, J. (2011). Epigenetics: Beyond face value. *Nature, 477*, 534–535.

West-Eberhard, M. J. (2003). *Developmental plasticity and evolution*. New York: Oxford University Press.

Wirtz, B. W., Mathieu, A., & Schilke, O. (2007). Strategy in high-velocity environments. *Long Range Planning, 40*(3), 295–313.

Witt, U. (2008). What is specific about evolutionary economics? *Journal of Evolutionary Economics, 18*, 547–575.

Witt, U. (2014). The future of evolutionary economics: Why the modalities of explanation matter. *Journal of Institutional Economics, 10*(4), 645–664.

Witt, U., & Cordes, C. (2007). Selection, learning and Schumpeterian dynamics: A conceptual debate. In H. Hanusch & A. Pyka (Eds.), *The Elgar Companion to Neo-Schumpeterian Economics* (pp. 316–328). Cheltenham: Elgar.

Zahra, S. A., & George, G. (2002). Absorptive capacity: A review, reconceptualization, and extension. *Academy of Management Review, 27*(2), 185–203.

Epigenetic Economics Dynamics in the Internet Ecosystem

Jon Mikel Zabala-Iturriagagoitia, Miguel Gómez-Uranga,
Jon Barrutia and Goio Etxebarria

1 Introduction

We are witnessing a very rapid development of mobile telephone-related sectors, firms, and technologies on the Internet. The changes being perceived at the present time are characterized by their high speed, giving rise to high velocity markets and high velocity environments (Eisenhardt and Martin 2000). In such environments, a number of dynamics are not explained by evolutionary principles. This chapter focuses on the dynamics observed in the Internet ecosystem as an illustration of the abrupt changes occurring in these high velocity environments (Fransman 2014).

Apart from well-known, established firms like Google, Apple, Facebook, Amazon, Samsung, Microsoft, IBM, Intel, Twitter, or Yahoo, new arrivals are increasingly occupying leading positions in the market. In particular, Chinese software firms like Baidu (the Chinese Google), Alibaba (the Chinese Amazon), Tencent (the Chinese Facebook), Weibo (the Chinese Twitter), and telecom providers like

J.M. Zabala-Iturriagagoitia (✉)
Deusto Business School, University of Deusto, Donostia-San Sebastian, Spain
e-mail: jmzabala@deusto.es

M. Gómez-Uranga · G. Etxebarria
Department of Applied Economics I, University of the Basque Country,
UPV/EHU, Bilbao, Spain
e-mail: miguel.gomez@ehu.es

G. Etxebarria
e-mail: goio.etxebarria@ehu.es

J. Barrutia
Department of Management and Business Economics,
University of the Basque Country, UPV/EHU, Bilbao, Spain
e-mail: jon.barrutia@ehu.es

© Springer International Publishing Switzerland 2016
M. Gómez-Uranga et al. (eds.), *Dynamics of Big Internet Industry Groups
and Future Trends*, DOI 10.1007/978-3-319-31147-0_3

China Mobile, Huawei, Xiaomi, or ZTE are challenging the global leaders. In this chapter, and following Miguel and Casado (see Chapter "GAFAnomy (Google, Amazon, Facebook and Apple): The Big Four and the b-ecosystem"), these few but large business groups will be described as the GAFAs (an acronym formed from the initials of Google, Apple, Facebook, Amazon, etc.) In short, the software ecosystem and the Internet ecosystem are particularly characterized as being small worlds led by few actors (Iyer et al. 2006). Most of these companies are very young, with a clear entrepreneurial origin. However, their short lives did not prevent them from being among the companies with the highest market capitalization at the end of 2014.

The fierce competition amongst these groups forces them to innovate in a continuous, systematic and particularly agile and quick manner. They are thus not only able to anticipate their rivals, but also to change—quasi endogenously—the ecosystem they compete in. Each company's staggering rhythms of innovation not only seek to change their own business ecosystem and the overall Internet ecosystem, but also to rapidly adapt their own ecosystem to the latter.

The goal of this chapter is to analyze the evolution and dynamics observed in the GAFAs, determining how these firms expand and diversify their activities, in what we refer to as epigenetic dynamics. Our aim is to understand how epigenetic dynamics of the business groups mentioned above address change in response to variations in their environments.

To a great extent, the evolution and speed of the dynamics in the Internet ecosystem are being affected and determined, among other factors, by[1]:

- An exponential reduction in the costs of means, infrastructures, equipment and tools used by users and developers, which together allow for improvements in delivery times and connection speeds.
- New materials that can replace those in use to date in key elements such as processors, devices or power supplies.
- The speed at which certain services such as mobile payment, e-commerce, etc., are being deployed on networks.
- Competition among multiple and diverse actors, especially among "big business groups," for the control of global markets and technologies. It is also worth highlighting the new race between large Internet companies (i.e., the GAFAs) on the one hand, and large telecom operators (e.g., Verizon, AT&T, Deutsche Telekom, Vodafone, Telefonica, and Orange) on the other.[2]
- The development of the Internet is also associated with certain challenges, which can compromise its potential advance and introduce uncertainties in relation to its development speed. In particular, we are referring here to the dangers associated to cyberattacks and the lack of security for information circulating on the Internet, which threaten each and every one of the actors engaged in the production and use of the Internet.

---

[1]See also Frey (2015).

[2]With respect to the above-mentioned determinants it is not possible to know in advance the speed and intensity with which they may occur.

However, it is not a matter of falling prey to fatalism that denies any reasonable progress, since organizations (public and private) can find ways to face the risks associated with attacks that afflict players participating in open networks. Following the Epigenetic Economic Dynamics (EED) approach described in Chapter "Introducing an Epigenetic Approach for the Study of Internet Industry Groups" of this book, all trends mentioned thus far will be explored as variations of the environment and the subsequent economic, social, political, ethical, etc., consequences, derived from the risks that individuals, business groups, and countries are plagued by.

We have discussed how adaptation to rapid changes in business environments gives rise to epigenetic dynamics. A large part of epigenetic dynamics is due to the groups' external relations, such as purchasing external assets, like small start-ups created by developers. Patents are the other key determinant of the epigenetic dynamics observed in the case of the GAFAs. This chapter on the dynamics of the large Internet industry groups therefore focuses on their patenting behavior and the mergers and acquisitions (M&As) completed.

The diversification paths followed by the GAFAs illustrate that most of these paths go against all odds or predictability. This makes prediction of the GAFAs' dynamics almost impossible ex-ante, because despite these business groups having an original and dominant specialization (i.e., DNA), they become active in areas that are unrelated to their original specialization. Accordingly, all groups become involved in the dynamics of the other groups, so "everyone does everything" and therefore "competes with everybody." This behavior finds its rationale in the fact that all the firms in the Internet ecosystem are interested in pursuing and positioning themselves in every dominant vector, because they cannot leave areas or segments to one side or unattended. Even if due to the specialization of each company their presence in every dominant vector may not be efficient from an economic point of view (i.e., they do not master the required skills or have the necessary competences internally), they must cover these segments to avoid being relegated or removed from the competition due to the dominance of the other actors in these segments. As a result, instead of talking about competing firms, large industrial groups on the Internet are regarded as "business ecosystems" (Moore 2005; Jing and Xiong-Jian 2011).

Patents are one of the aspects that best characterizes the epigenetic dynamics of the Internet business groups we address in this chapter.[3] The GAFAs need to acquire large portfolios of thousands of patents not only to protect those areas that represent their core capabilities (e.g. Google with its Android and the acquisition of Motorola's patent portfolio), but also to gain access to those areas where they do not have particular strengths. Such convulsive patent purchasing is known as the patent war (Carrier 2012; Paik and Zhu 2013; Lim 2014; Cass 2015), which is particularly relevant in the context of the GAFAs (Gómez-Uranga et al. 2014). Thus, an exacerbated patenting behavior can be observed, particularly as regards the increase in the number of patent applications (voluminosity) in recent years

---

[3]We are very much aware of the bias in using patents as a proxy for innovativeness. However, given the remarkably high patenting in the industry, we believe this indicator can provide an accurate picture of the dynamics within it.

(van Zeebroeck et al. 2009). The same outcome can also be observed regarding the acquisition of patent portfolios and business groups that initially were not involved in areas related to the Internet ecosystem (e.g., health).

The purpose of this chapter is to analyze the extent to which the GAFAs' dynamics are more or less disruptive, and therefore closer to or further from the original DNA of these groups. In this regard, we pose the following hypothesis in relation to the EED approach outlined in Chapter "Introducing an Epigenetic Approach for the Study of Internet Industry Groups": the further from a given branch of activity a firm is, the more disruptive their epigenetic dynamics will necessarily be, as this distance forces the firm to take fast and radical measures to get into these new branches. To test this hypothesis, we first describe some of the GAFAs' dynamics during the last 3 years (2012–2015). We then analyze whether the observed epigenetic dynamics can be explained by the patenting behavior of these business groups. To achieve this goal, we study the patents granted to these groups at the USPTO. Our study is focused on the USPTO instead of on the EPO or the PCT patents because the patenting of software, which is one of the central elements that helps characterize and explain the GAFAs' evolution, is permitted in the US (Chingale 2015; Useche 2015). Our aim is thus to analyze the behavior of the groups in their patent portfolios to illustrate the EED approach. In this way, we aim to provide an answer to the following research questions (RQ) in relation to patents:

- RQ1: What technological areas are the GAFAs moving into over time? Are they related to these groups' DNAs?
- RQ2: Are these diversification dynamics similar across the GAFAs?

The next aspect that we focus on in order to explain the epigenetic dynamics identified is that of the M&As carried out by these groups. In this regard, the number of M&As completed by each of the GAFAs, and the amount of investment required are discussed. As with the case of patents, in the case of M&As we also aim to analyze whether the newly acquired companies are related to the original and dominant specialization of each business group or if, on the contrary, the M&As are carried out in a defensive mode so as to guarantee a fast adaptation to the new environment. The research questions we aim to address in relation to the M&As can be formulated in the following way:

- RQ3: What is the amount of investment required?
- RQ4: Are there different strategies among the firms or do they follow similar paths?

The relevance of external sources of knowledge as determinants of innovation has been emphasized in the literature from a wide variety of approaches. The literature on innovation systems highlights innovation as being the result of dynamic social and economic processes based on learning and interaction among actors (Lundvall 1992). Network theories (Håkansson 1987) also maintain that companies rarely innovate individually and that the introduction of new products or processes in the market depends on their ability to cooperate with external agents. Similar arguments are also posed from strategic management perspectives, which note that the search for new ideas, new organizational forms, etc., surpasses the boundaries of the organization (March 1991). Chesbrough (2003) describes this phenomenon as the rise of open innovation modes (Huizingh 2011; Mortara and Minshall 2011; van de Brande et al. 2009).

However, the concept of absorptive capacity (Cohen and Levinthal 1990; Zahra and George 2002; Engelen et al. 2014) suggests that internal capacities can only be improved in the knowledge bases that are relatively close to those available into the firm. In other words, a company will be able to exploit external knowledge as long as this knowledge can be identified and assimilated. This has led some scholars to introduce the concept of "related variety" (Frenken et al. 2007). As Asheim et al. (2011) discuss related variety refers to a set of complementary sectors that share capabilities and competences so that it becomes easier to understand and absorb each other's knowledge. The underlying idea is that "a region specializing in a particular composition of complementary sectors will experience higher growth rates than a region specializing in sectors that do not complement each other" (Frenken et al. 2007: 686). Consequently, related variety-driven firms, entrepreneurial ventures, territories, etc., reduce the risk of selecting "wrong" activities since the existing competences are taken as the point of departure in order to broaden the economic base.

Specialization in certain fields tends to increase the risks associated to potential external shocks due to a lack of diversity. In contrast, it can also be argued that the wider the sectoral variety, the higher the probability of promoting economic growth. Accordingly, it should also be noted that although these variety-driven spillovers can lead to risk reduction, they may also reduce the probability of obtaining higher profits. This is what is often known as the "unrelated variety" phenomenon. According to this, when one sector is hit by an economic downturn in a territory with a high degree of unrelated variety, this will not negatively affect the other sectors. In sum, while unrelated variety safeguards against external shocks, related variety is expected to be beneficial for Jacobs externalities in the form of knowledge spillovers (ibid: 688).

Several studies have been conducted on these two concepts (see Parrilli and Zabala-Iturriagagoitia 2014), studying whether territories specialized in certain activities, industries or products (i.e., where related variety is in place) are more conducive to innovation and growth as compared to other locations that have more diversified industrial structures (i.e., unrelated variety). In this case, this chapter examines the diversification paths followed by the GAFAs through an analysis of their patenting strategies and their M&As, in order to determine whether their strategies are related or unrelated varieties.

Finally, and following the methodology outlined in Chapter "Introducing an Epigenetic Approach for the Study of Internet Industry Groups" in this book, we also analyze the consequences of these epigenetic dynamics. This analysis of consequences is mainly limited to the patent system, although other dimensions are also discussed. Intellectual property has become one of the main pillars of the economic dynamics of Internet industry groups. Patent lawsuits for infringements and violations are quite common (Cunningham 2011). The biggest disputes have taken place in the mobile phone business and have even affected consumers. In turn, these lawsuits create a large number of ad hoc alliances out of sheer self-interest, thus distorting the original defining features of each group and, above all, hindering possible innovations, competition and progress for users. Sometimes companies seriously damage the competition with lawsuits, asking courts to stop the sale of their rivals' products. In this regard, a distortion in the rationale for patenting, excessive transaction (and litigation) costs, high entry barriers to SME patenting, problems in the definition and development of standards and an overload in patent offices and regulating agencies are regarded as the main consequences or malfunctions.

The large amounts of financial and human resources that must be devoted to patenting and patenting litigations also become key entry barriers for small developers and start-ups. All these consequences ultimately create disincentives or anomalies for many innovative firms and entrepreneurs, which contrasts with the rationale of the patenting system for supporting the development of further innovations.

The rest of this chapter is structured as follows. Section 2 introduces the concept of business ecosystem and its usefulness in understanding the dynamics of the large Internet industry groups. In addition, it discusses some of the characteristics that define the Internet ecosystem. Section 3 provides some of the main structural characteristics of the firms under study, and illustrates the business ecosystems of the GAFAs under study. Section 4 provides a preliminary introduction to the epigenetic dynamics that we have observed in the evolution of the GAFAs. It helps understand their expansive strategies, the new industry segments they are active in and how patents and M&As become essential in explaining these moves. Section 5 provides the main patenting dynamics that are observed, while Sect. 6 does so with the M&As. In both cases, firms are analyzed separately since each of them covers different periods due to the differences in their year of constitution. Understanding what these consequences are opens the way for the definition of more effective policies in a variety of domains, such as science, technology, and innovation policy (both from the side of supply and demand-side instruments), industrial policy, entrepreneurship policy, employment generation, regional development, education, the institutional environment (i.e., regulations), the patent system, etc. Finally, Sect. 7 concludes by discussing some of the most relevant consequences of the previous epigenetic dynamics.

2 The Internet as a Business Ecosystem

The EED approach is the main focus of this book. However, in this chapter the study of the GAFAs' dynamics is approached not only from an epigenetic perspective but also from that of the business ecosystem. In this way, we intend to bring together two streams of literature (one proceeding from molecular biology and the other from industrial systems) that each have their own logic (Herstatt and Kalogerakis 2005; Cordes 2006).

Referring to the main differences between biological and industrial ecosystems, Daidj (2011) highlights that the evolution of a business ecosystem depends on the decisions made by firms and the institutional and regulatory framework around them. These decisions are intentionally made by the firms themselves,[4] in the sense that they use them to achieve a certain goal or move in a certain direction. However, such intentionality is not found in biological ecosystems.

The business ecosystem approach helps us explain how business groups are being structured with regard to their respective stakeholder groups (e.g., customers, users, suppliers, investors, institutions, regulatory authorities, standard-setting bodies, etc.), the relationships with these and the overlaps and/or conflicts that may emerge with other business ecosystems (Teece 2007). However, it does not make it possible to reach a comprehensive understanding of their evolution, although like any ecosystem they are undergoing continuous evolution (Gueguen and Isckia 2011). To address this gap, we rely on the use of the EED approach. Therefore, in this chapter, we intend to make the two approaches complementary, as they allow us to study the same units of analysis but from two different angles.

But what is an ecosystem? There are many scholars who have tried to incorporate the concept of ecosystems into different streams of the literature (Bijker et al. 2012; Autio and Thomas 2013; Clarysse et al. 2014). The concept of industrial ecosystem was first introduced by Frosch and Gallopoulos (1989). From their view, an industrial ecosystem functions as an analog of biological ecosystems. Their reflection dealt with the impact and consequences of technology on industry and society at large. Their main interest revolved around the optimum consumption of energy and materials, minimization of waste generation and the maximization of material reutilization (i.e., recycling).

The economy as an ecosystem was first introduced by Rothschild (1995), for whom a capitalist economy should be understood as a living ecosystem in which competition, specialization, cooperation and growth should be regarded as determinants of economic behavior. Rothschild equated firms with biological organisms and industries with species. Firms and industries also have their own genes, which, in part, determine their behavior; and, like living organisms, firms also relate to other actors in the ecosystem, where other predators and prey also coexist.

The concept of business ecosystem was first suggested by Moore (1996), who considered that biological metaphors could be useful when understanding economic and industrial processes. Moore initially defined a business ecosystem as an

---

[4]We will not engage in a discussion here on whether these decisions are voluntarily made by the firms because they are part of their strategic reflections or because the environment somehow "forces" them to do so. Nor will we focus on how these decisions are made, namely: individually or collectively; fast or slowly; following systematic, routinized and structured processes, or chaotic and improvised ones instead; and adopting a heuristic approach or a strategic one, in which information is gathered to form a comprehensive picture of the environment and the firm's relative competitive position in it (Bourgeois and Eisenhardt 1988; Eisenhardt 1989, 2001, 2013; Judge and Zeithaml 1992; Clark and Collins 2002; Hernández-Martínez 2006; Bingham et al. 2007; Christiansen and Varnes 2007; Davis et al. 2009; Hadida et al. 2015; Palmié et al. 2015).

economic community based on a set of interacting individuals and organizations (Moore 1996: 26). In his work, market leaders were characterized as keystones that had a fundamental influence on the coevolution processes of the ecosystem (also see Jansen et al. 2013). However, in his later work he moved on to consider many other types of stakeholders, which were considered "mutually support- ive organizations" as agents that could be encompassed within a single business ecosystem: "Business ecosystems are communities of customers, suppliers, lead producers, and other stakeholders—interacting with one another to produce goods and services. We should also include in the business ecosystem those who provide financing, as well as relevant trade associations, standards bodies, labor unions, governmental and quasigovernmental institutions, and other interested parties. These communities come together in a partially intentional, highly self-organiz- ing, and even somewhat accidental manner. But the result is that the members pro- vide contributions that fill out and complement those of the others" (Moore 1998: 1681–69). Finally, Moore (2005) considered that the agents in an ecosystem and the decisions they make are influenced by those of all the other actors in the same ecosystem. Business ecosystems would thus "refer to intentional communities of economic actors whose individual business activities share in some large measure the fate of the whole community" (Moore 2005: 3). In this manner, the members of an ecosystem may form a broad system of organisations that support each other mutually: communities of customers, suppliers, leading producers, business asso- ciations, standardization bodies, etc., which are interested in joint work and coop- eration for the good of the community. This is the way in which the concept of business ecosystem is understood in this chapter.

For Iansiti and Levien (2004), the concept of business ecosystem is worth con- sidering for analysis of the structure of large business groups, because, as in the case of biology, in the economy there is also a wide variety of agents (i.e., individ- uals, organizations, institutions) in continuous interaction and which are mutually dependent for their respective survival. As a result, Iansiti and Levien discuss how robust organizations are and whether or not these are able to adapt, and therefore survive, vis a vis internal and/or external changes.

According to Iansiti and Richards (2006) and Gueguen and Isckia (2011), the concept of ecosystem reaches its maximum expression in ecosystems such as those of Apple, Amazon, Google, or Facebook.[5] These groups are characterized by their high and rapid innovativeness and the large number of actors that are embed- ded in their respective ecosystems. In this regard, after a careful review of the liter- ature, Autio and Thomas (2013: 205) conclude that an innovation ecosystem can be defined as "a network of interconnected organizations, connected to a focal firm or a platform, that incorporates both production and use side participants and cre- ates and appropriates new value through innovation."

---

[5]Autio and Thomas (2013: 208) also consider the adequacy of the concept of ecosystem when applied to organizations (i.e. hubs) like eBay or platforms like Android.

The concept of business ecosystem is increasingly being applied in the context of the Internet (Nachira 2002; Fransman 2014). Numerous studies can be found in the framework of business ecosystems (Corallo et al. 2007). To mention a few, Razavi et al. (2010) discuss how digital business ecosystems are based on loosely coupled interactions and have demand-driven properties and self-organizing characteristics, which may help small and entrepreneurial firms to constitute new networks and/or become embedded in already existing ones. A similar approach is followed by Ndou et al. (2010), who introduced a dynamic integrated focus based on the modularity concept, which shows several value creation potentialities within the context of digital business ecosystems. Karakas (2009) also considers that the web has become a digital ecosystem, which is characterized by creativity, community, connectivity, cooperation, and convergence.

Moore (2005) proposes that companies manufacturing Apple's iPod belong to the iPod business ecosystem. Basole (2009) analyses the interrelationship between device manufacturers, telephone operators, and platforms (Symbian and Windows). Jing and Xiong-Jian (2011) analyze the strategies of mobile network operators in China. Wan et al. (2011) approach the analysis of the Chinese software ecosystem according to stability factors (e.g., diversity, resilience) and sustainability factors (productivity, vitality, creativity). Similarly, based on the Italian context, Battistella et al. (2013) introduced a model to systematically study the structure and fluxes of the networks in a business ecosystem. Han and Park (2010) have mapped the relationships between the technology, products, and services they produce. Li (2009) explains how Cisco Systems has deployed an intentional M&A strategy to sustain its corporate growth, by means of the analysis of Cisco's technological roadmap according to their patents in the USPTO.

The business ecosystem approach assumes that there is in fact interdependence between the agents within an ecosystem (Adner and Kapoor 2010). However, the role and importance of each one varies, not only over time, but also within the multiple projects that may be underway at the same time within a particular ecosystem (Fig. 1). As an example, application developers are found in the area of software, and although some work with Android, others interact with Apple, others with Symbian, and many with all of them at the same time, meaning competition takes place in different ways in the ecosystem (OECD 2013). In other words, companies compete and cooperate with each other within an ecosystem (Gueguen and Isckia 2011).

Changes in ecosystems come from changes within, although they are influenced and sometimes determined, as in epigenetics, by what is happening around them (i.e., selection environment). Setting clear boundaries on where a certain ecosystem ends is a hard task due to their openness and permeability. The defining element of an innovation ecosystem is not a given product, but rather a coherent set of interrelated technologies and associated organizational competences that bind a variety of participants together to coproduce a set of offerings for different user groups and uses (Autio and Thomas 2013: 208; Geels 2014).

Competition and collaboration sometimes coexist, so it is often difficult to separate the two. Applications and content from a multitude of providers are supplied

Fig. 1 The map of the Internet. *Source* http://orig01.deviantart.net/91af/f/2014/070/a/5/map_of_
the_internet_2_0__by_jaysimons-d781bst.jpg. Accessed 19 December 2015

through most of the GAFAs' platforms (e.g., their APIs[6]). These platforms are
understood as the coordinating artifact that the firm at the center of the business
ecosystem (i.e., each of the GAFAs) "uses, or the services, tools, and technologies
that other members of the ecosystem can use to enhance their own performance"
(Autio and Thomas 2013: 208). Platforms are also often associated with network
or spillover effects, so that the more users adopt the platform, the more valuable
this becomes to the owner and to the users, because of growing access to the network of users and often to a set of complementary innovations (Gawer and
Cusumano 2012: 1; Gueguen and Isckia 2011).[7] Logic, in terms of market size,
also merits a closer look in the case of the GAFAs. As a greater number of applications become available in their platforms, developers show a greater interest in
creating new ones and the GAFAs in providing them. These applications form the
foundations of the GAFAs' ecosystems because they are constantly increasing the
platform's usefulness and value.

---

[6]In computer programming, an Application Programming Interface (API) is a set of routines,
protocols and tools for building software applications. An API expresses a software component
in terms of its operations, inputs, outputs, and underlying types. An API defines functionalities
that are independent of their respective implementations, which allows definitions and implementations to vary without compromising the interface. A good API makes it easier to develop
a program by providing all the building blocks. A programmer then puts the blocks together
(Wikipedia, 2015—https://en.wikipedia.org/wiki/Application_programming_interface).

[7]In biology, these phenomena in which some species' actions help others is referred to as
mutualism.

Software and application developers, for example, can be considered collaborators since they are embedded in the ecosystems of most GAFAs. They do not exclusively provide content for Apple, but also offer their services through other systems such as Google's Android. Competition in the ecosystem is thus not divided into separate activities but takes place within the whole, meaning that if another group wanted to compete with Apple (or any of the GAFAs), it would not only have to offer a better device than the iPhone, iTunes or the iCloud (the products/services of the other GAFAs), but an entire ecosystem of applications and content. Eaton et al. (2011) consider that in the particular case of Apple, the radical innovation that was produced with the introduction of the iPhone cannot only be explained by the device itself or the applications that can be installed in it, but rather by the platform around the phone itself, which is cocreated by Apple, the developers and other stakeholders.[8]

In biology, the concept of coevolution is used to illustrate the reciprocity of changes (Peltoniemi and Vuori 2004). Coevolution thus implies that when changes in a certain species are produced, these also induce changes in others (i.e., variation). Besides coevolution, other terms such as coexistence, predation, symbiosis, or parasitism could also be used to illustrate this phenomenon (i.e., in evolutionary terms we would talk about variation, retention and selection, changing routines, learning, developing absorptive capacity, and dynamic capabilities). In fact, for Gueguen et al. (2006), when a large firm acquires another and integrates it within its structure (i.e., vertical integration), this could be seen as a symbiotic or predatory process. Another example of symbiosis can be seen in Facebook's partnership with Netflix or Spotify.

Platforms, which support collaboration and relationships, are the mechanism that makes it possible to enlarge and organize an ecosystem. Platforms appear in different ways in information and communication ecosystems, whether they are operating systems (e.g., Microsoft, Linux), APIs (e.g., Google, Facebook) or the like. As we discussed earlier, even devices such as the iPhone or iPad could be considered platforms due to the multiple activities, actions, and services that can be used on them.

2.1 Characterizing the Internet Ecosystem

Having discussed some of the different views on ecosystems, our intention is now to stress some of the underlying features that help characterize the Internet as a business ecosystem (also see Rong et al. 2015).

(a) Competition takes place among ecosystems

During the past two decades of the twentieth century, vertical integration began to characterize the dynamics of large business groups globally. This increased the

---

[8]Eaton et al. (2011: 2) use the term generativity, which refers to "the ability of a self-contained system to create, generate, or produce new content, structure, or behavior without additional help or input from the original creators".

complexity of their strategies, because their movements (i.e., acquisitions, mergers, takeovers or alliances) could not be analyzed from a single point of view. In the case of GAFAs, their strategies are also becoming increasingly complex because different movements coexist in space and time, making it impossible to establish regulations a priori, as vertical integration strategies can only be analyzed retrospectively and on a case-by-case basis. This also applies to Internet regulation and hinders the implementation of industrial or competition policies.

The analysis of vertical integration already considered the existence of different industries that were related to the original activity of the company or group that was seeking integration (i.e., related variety). The concept of ecosystem allows competition to be understood in a more comprehensive way, as it makes it possible to consider not only the existing concentration in a specific industry but in the ecosystem. The fact that ecosystems are continuously undergoing rapid growth makes it impossible to control the whole ecosystem, leading to systemic failures that have a significant economic and social impact.

(b) The layers of the Internet ecosystem

As Barua et al. (1999) observe, the Internet economy can be characterized by its four-layer model (Table 1). Barua et al. (1999) first classify the Internet economy into two broad categories: infrastructure and economic activity. The infrastructure category is then broken down further into two distinct but related layers. The Internet infrastructure layer provides the physical infrastructure for electronic commerce, while the applications infrastructure layer includes software applications, consultancy, training, and integration services. In turn, the economic activity category is divided into two other layers: online transaction and electronic intermediaries. The transaction layer involves the ability to guarantee the development of direct transactions between buyers and sellers. Finally, the intermediary layer involves a variety of parties providing capacities such as certification, search and retrieval of services that reduce transaction costs, etc. (ibid).

The GAFAs are mainly located in layers 3 and 4. However, much of the software they use to manage their databases (e.g., Amazon) or to build-up their software (e.g., Google), together with the central role of the network infrastructure, are activities located in layer 2.

(c) The coexistence of competition and collaboration

As we discussed earlier, it is sometimes hard to draw a clear line between cooperation and competition when considering the GAFAs' ecosystems (Brandenburger and Nalebuff 1996; Daidj 2011; Gueguen and Isckia 2011). For example, applications (i.e., apps) and contents made by a huge range of providers (i.e., from individual developers to large firms) are offered through the iTunes platform by Apple or the Google Play platform.[9] These content providers (e.g., King, Gameloft, Electronic Arts, Rovio, Disney, Supercell, Tencent, Line), as part of the Apple

---

[9]For a review of the top app trends, see App Annie (2015).

Table 1 A conceptualization of the Internet economy

Category 1: Infrastructure		
Layer 1: Internet infrastructure	Includes companies with products and services that help create an IP-based network infrastructure	– Internet backbone providers – Internet service providers – IP Networking hardware and software companies – PC and Server manufacturers – Fiber optics – Line acceleration hardware manufacturers
Layer 2: Internet applications infrastructure	In addition to software applications includes the human capital involved in the deployment of e-commerce and e-business applications	– Internet consultants – Internet commerce applications – Multimedia applications – Web development software – Search engine software – Online training – Web-enabled databases – Security products and services
Category 2: Economic activity		
Layer 3: Transactions	Increase the efficiency of electronic markets by facilitating the meeting and interaction of buyers and sellers	– Market makers in vertical industries – Online travel agents – Online brokerages – Content aggregators – Portals/Content providers – Internet ad brokers – Online advertising
Layer 4: Electronic intermediaries	Involves the sales of products and services to consumers or businesses	– E-tailers – Manufacturers selling online – Fee/Subscription-based companies – Airlines selling online tickets – Online entertainment and professional services

Source Adapted from Barua et al. (1999)

ecosystem, could be regarded as Apple's partners, or elements providing the ecosystem with a large variety. However, these same partners are not offering their products/services exclusively through Apple's platform, but also on many others like Android or Facebook.[10] For example, in the particular case of the media and music industries, the number of suppliers is very limited, as it is a highly consolidated segment. Thus, these large actors are not interested in offering their products exclusively to one of these platforms, which would imply that a particular ecosystem (e.g., Apple) sells the content of providers A and B, and another (e.g., Amazon) sells that of C and D. In fact, the opposite is the case.

This leads us to introduce what might be referred to as the "perimeter of innovation." Google, or any of the GAFAs, can develop their own hardware, software,

---

[10]Spencer (2015) provides evidence of the number of apps developed by Google, Facebook, Microsoft and Adobe, which are provided through the Apple's iOS app store.

APIs, apps, platforms, etc., in-house, taking into consideration the inputs provided by their strategic partners, and/or organizing and promoting events (e.g., hackathons, Kickstarter competitions, Techcrunch events, web summits, code events, Startup battlefields) they may consider strategic for such purposes. However, as more organizations and actors are being embedded in the respective ecosystems, the management of their innovation processes becomes more complex, as the GAFAs need to show the ability to develop and maintain multiple partnerships at the same time with content providers, app developers, etc., each of which also has their own innovation strategies. By the term perimeter we mean that it is often not possible to manage the entire ecosystem and the multiple innovation processes/projects co-existing within it (i.e., retention). Each actor within the ecosystem has its own timing, pace and orientation. Thus, as we move toward the perimeter of the ecosystem, that is, as we draw away from the center of the ecosystem (i.e., the DNA of the firm), managing the innovation and the required partnerships becomes more complex.

(d) The leadership and the dominant vectors

As Autio and Thomas (2013: 208) observe, "while the number of digital services grow in a linear fashion, the distribution of complementors to hub firms [i.e. the GAFAs] tends to follow a power law, implying that a small number of hub firms provided for a majority of complementors." As we mentioned in the introduction, the software ecosystem, and the Internet ecosystem in particular, are characterized by being small worlds led by few actors (Iyer et al. 2006). The leaders (i.e., "keystone organizations" in the words of Iansiti and Levien 2004 or Jansen et al. 2013) in each dominant vector can have a relevant influence not only on their ecosystem but also on that of their competitors (Pellegrin-Boucher and Gueguen 2004; Moore 2005). A change in Apple's ecosystem may induce changes in the rest of the big Internet industry groups. Each of these business groups is a leader in its main field, which means they set the patterns for others, although this is mutual. Google+ copied Facebook, which in turn took many ideas from Google+. Apple first supplied content download and later integrated the applications. In each case, these activities were initially complementary, which gave Apple a slight edge. These business moves were later copied, to varying degrees, by the rest of the groups.

Despite all the big players in the Internet ecosystem being forced to perform the same activities, as already indicated, the original DNA in each group means that each has a leading position in particular dominant vector. As a result, some of the activities performed by each of these business groups are more central to their evolutionary dynamics than to those of others. This can be either because these activities are those central to its own ecosystem or because they provide the group with some kind of competitive advantage. In the case of Apple, for example, it could be considered that its leadership is in its original DNA, namely, the design and manufacture of hardware and software. However, having its own platform for downloading content and applications like iTunes and a platform for contents and software on the cloud like iCloud, which allows users to keep all synchronized content available at any time, gives it a competitive advantage over the rest. On the other hand, the fact that certain activities are not central to the other groups means that they need to carry out certain dynamics in order to follow the patterns established by the previous group. As previously discussed, the further from a given branch of activity a firm is, the more disruptive its epigenetic dynamics will necessarily be, as this distance forces the firm to take fast and radical measures to enter these new branches. Subsequently, as we will demonstrate in Sect. 4 of this chapter, the leaders transmit their DNA in the Internet ecosystem through the activities they perform, so that these are copied, to a greater or lesser extent, through epigenetic moves made by the rest of the GAFAs (i.e., related to the retention and selection mechanisms in the evolutionary realm).

(e) The environment

As explained earlier in the book (see Chapter "Introducing an Epigenetic Approach for the Study of Internet Industry Groups"), the environment is one of the key elements for understanding epigenetic dynamics. Like genetic contexts, the DNA can be a central element in explaining certain diseases, although epigenetic factors are often as important, or even more so. In the context of the Internet ecosystem, one of the determining aspects of the environment are the institutions around it, which are increasing the number of regulations that drive the dynamics and evolution of the GAFAs and their respective ecosystems. One example of these institutions is the decisions that different competition courts have handed down in the USA and Europe concerning the mergers and acquisitions of some of these groups. Other more recent examples could be those related to network neutrality, which may modify some of the current practices of the GAFAs and may even hinder the survival of some of them in the medium term. Finally, we can also refer to the regulations that frame the activities of certain industries. For example, the provision of physical infrastructures by telecom providers (see Chapters "Future Paths of Evolution in the Digital Ecosystem" and "4G Technology: The Role of Telecom Carriers" in this book) is a key determinant of the current dynamics of the Internet ecosystem and of the GAFAs in particular. However, the provision of telecom services is a highly regulated industry while, to date, the Internet has lacked (almost) any regulation, or it has not even been possible to regulate it. As the 2013 report to the United States Securities and Exchange Commission on Twitter states "… these laws and regulations may involve privacy, rights of publicity, data protection, content regulation, intellectual property, competition, protection of minors, consumer protection, taxation… many of these laws and regulations are still evolving and being tested in courts and could be interpreted in ways that could harm our business. In addition, the application and interpretation of these laws and regulations are often uncertain, particularly in the new and rapidly evolving industry in which we operate" (SEC 2013). As the penetration of the GAFAs into other (often unrelated) industries is becoming increasingly consistent over time as a result of their expansive and defensive strategies, it is possible to observe how many of these are moves into industries that are very highly regulated, such as the provision of telecom services, the health industry, human mobility, education or banking. In Sect. 4 of this chapter, we examine these moves more closely, which, as indicated, in the context of this book we refer to as epigenetic dynamics, due to the fact that they frequently have no relationship with the business groups' original DNA.

The increasing number of reflections on the role use and often misuse of copyrights, and particularly patents, could also be included in these environmental institutions (see Sect. 5 in this chapter). Despite the number of devices that are constantly being connected to the Internet and the continuous increase in Internet traffic (which may increase even further with the advent of the Internet of things), these go hand in hand with a reduction in the openness of the Internet. In fact, the increasing number of apps could be regarded as activities that skip the "open" Internet searches as we have known them to date. As a result, in the short-term, online Internet search engines may become entirely different from today's; with such changes including app searches and allowing apps to be installed in the corresponding devices. However, it might also be the case that the app developers themselves could be interested in providing this app search service. Social networks also provide a good example of the reduction in the open space of the Internet, as their content is often not included in open searches. Here, too, the decision about which platform to use very much depends on the number of contacts, information and applications that can be accessed within them.

(f) Phenotypic and genotypic changes and the impossibility of making predictions

In Chapter "Introducing an Epigenetic Approach for the Study of Internet Industry Groups" of this book, we introduced the differences between the genotype and the phenotype. We discussed how the plasticity of the genome enables it to adapt to the environment resulting in the formation of different phenotypes determined by the environment the organism is exposed to.

The ability to make predictions, which we are so used to (particularly in economics), on the future development of firms, ecosystems, countries, etc., is only possible in those cases where the variables considered conform to certain previously established and known guidelines and patterns. This would be the case of disciplines such as micro and macroeconomics. An example of this logic is provided by many of the reports that some of the most well-known consulting firms and supranational bodies prepare with a certain periodicity, and which often contribute to expanding the notion that the future is, to a great extent known and predictable. In the case of the Internet ecosystem in particular, the Global Entertainment and Media Outlook by Pricewaterhouse Coopers,[11] the Internet of Things Outlook report,[12] the OECD Digital Economy Outlook,[13] the report on the

---

[11]The Global Entertainment and Media Outlook 2015–2019 is available: http://www.pwc.com /gx/en/global-entertainment-media-outlook/. Accessed 15 August 2015.

[12]The Internet of Things Outlook report 2015 is available: http://telecoms.com/intelligence/iot-outlook-2015/. Accessed 15 August 2015.

[13]The OECD digital economy outlook 2015 is available: http://www.oecd.org/internet/oecd-digital-economy-outlook-2015-9789264232440-en.htm. Accessed 15 August 2015.

future of the Internet by the World Economic Forum,[14] or the IPG Media Lab' Outlook Report[15] could be highlighted.

However, in the Internet ecosystem, as is also the case in many other high-velocity environments, prediction loses much of its meaning, since it is clear that the rates of change are proving to be much faster than originally expected. Using a metaphor, we could think of the GAFAs as a landscape of dunes, changing constantly, hence capturing their features in snapshots would not make much sense. We therefore agree that the dynamics we intend to observe in the GAFAs in this chapter are far more complex than what we are capable of studying. However, with this book in general, and this chapter in particular, we want to take a first step in this direction. One of the defining characteristics of every ecosystem, like any living organism, is its ability to adapt and evolve (Basole 2009), either due to genetic evolution or to epigenetic transformation. Some of these recent changes, the consequences of which are still unknown, could be as follows:

- Growth of the immaterial (software, content, platforms): this does not imply that hardware and physical infrastructures are forgotten. In fact, software is being increasingly embedded in them.
- Technical convergence: different types of services (e.g., voice, text, code, data, image, video, or a mix of them) are bypassing the physical infrastructure (i.e., cables) available to date, which is also undergoing very large changes (see Chapter "4G Technology: The Role of Telecom Carriers" on telecom carriers and the deployment of 4G technologies and infrastructures). In turn, these services can pass through multiple devices (e.g., PC, laptop, smartphones, iPad, online TV, etc.) and multiple distribution and transmission technologies (e.g., satellite, air, cable, etc.).
- Functional convergence: for example, in the information contained in an online newspaper there is a convergence of text, images, videos, social networks, etc. In turn, televisions use the same basic information, so the different media are increasingly looking more alike and competing not only among themselves, but also with the GAFAs (i.e., Google news, Yahoo news, etc.).
- Corporate convergence: nowadays the same companies are operating in sectors that were previously separate. Thus, a firm participates in many industries, and the boundaries across sectors are becoming increasingly blurred (Gueguen and Isckia 2011).
- Ecosystemic convergence: this last type of convergence is related to the fact that, as we will see in the course of this chapter, each of the GAFAs is increasingly integrating more activities that are not related to its core or original DNA but rather to the original activities of its competitors.

---

[14]The report on the future of the Internet is available: http://reports.weforum.org/outlook-global-agenda-2015/future-agenda/mapping-the-future-the-future-of-the-internet/. Accessed 15 August 2015.

[15]The IPG Media Lab' Outlook Report 2015 is available: http://ipglab.com/outlook2015/. Accessed 15 August 2015.

(g) The consequentialist change

Changes are evolutionary and constant, influencing not only further development of the business ecosystem itself but also having an indirect impact on the dynamics and paths followed by other ecosystems which, initially, are not directly related to the GAFAs' ecosystems. This has become evident recently, as the actions undertaken by the GAFAs are having a direct influence on other "offline" industries, such as taxis, accommodation, mobility, retailing, health, etc. (Evans et al. 2008).

It can be said that the dynamics of the business ecosystems (concerning the GAFAs) are endogenous to and consciously developed within the Internet ecosystem. The leaders in the respective dominant vectors develop their strategies, involving other stakeholders in their respective ecosystems, and the competing firms are obliged to follow those paths, requiring that they undergo disruptive dynamics due to the changes in the environment. It could even be said that the dynamics of the Internet ecosystem are produced in response to the changes within each of the business ecosystems of which it is formed (i.e., genetic evolution in a Darwinian sense). Each of these business ecosystems is also influenced by its respective (direct) environments (i.e., epigenetic evolution). However, it also has a direct impact on the behavior of the latter (Breslin 2011; Breslin et al. 2015), in the sense that the dynamics of the business groups and their ecosystems also constitute the dynamics of the global Internet ecosystem as a whole. In this regard, when dealing with a high velocity market such as that observed on the Internet, it is not enough to consider the evolution of firms as mere adaptation in response to the changing conditions of the environment (first step of the EED approach). In addition, the dynamics of the firms in the Internet ecosystem should also be taken into consideration (second step of the EED approach), as their strategies are often not only seeking to survive in the selection environment, but also to move ahead of the changes observed in it by making fast and disruptive moves. But even then, in consequentialist logic, they have an indirect impact on the performance and evolution of many other ecosystems; a global impact that has economic, social, institutional, regulatory and even moral consequences (third stage of the EED approach). Thus, with the methodology outlined in Chapter "Introducing an Epigenetic Approach for the Study of Internet Industry Groups," we aim to provide a comprehensive view of the trends and dynamics observed in the Internet ecosystem.

3 Identifying the Genomic Instructions of the GAFAs

The goal of this section is to characterize the business groups we are interested in exploring. Applying the three-stage EED approach outlined in Chapter "Introducing an Epigenetic Approach for the Study of Internet Industry Groups," in this section we identify the business groups' original DNA. Namely, their original genomic instructions and the products/services that characterized their constitution and consolidation (i.e., the first stage of the EED approach). The first innovative products that are most closely identified with the business groups' initial activities are those forming the essence of each company (e.g., Google's search engine, Microsoft's operating system (OS), Apple's software and design, the social network concept developed by Facebook). Next, the focus moves to the structural characteristics of the ecosystem in which these firms operate and which may lead to changes in their DNA. Some of these characteristics are: intense inter-group competition, exponential growth of markets and users, high demand for innovation, expansion in the number of applications and their content, fast multivectorial technological change and planned obsolescence, modularity in the behavior of business ecosystems, etc. The previous dimensions are due to the high variability of these big groups' environments (Wirtz et al. 2007). Then, in Sect. 4, we move on to the second stage of the EED approach, where an analysis of their epigenetic dynamics is undertaken. Thus, with this section we aim to frame the overall structural characteristics of the GAFAs. In this chapter, the firms that we consider within our group of GAFAs are the following: Google, Apple, Facebook, Amazon, Samsung Electronics, Microsoft, Nokia, Twitter, eBay, and Yahoo.

Before starting with the characterization of the GAFAs, two issues should be noted. One of them is rather simple or even anecdotal, while the other is essential. The first is that there seems to be no other industry that is followed to the same extent by users themselves, followers and the media in general. The names of Bill Gates (Microsoft), Steve Jobs (Apple), Mark Zuckerberg (Facebook), Larry Page (Google), Jeff Bezos (Amazon), etc., are probably familiar to everyone. However, this cannot be said for most of the other major companies worldwide. The second point to stress is that we are currently witnessing many changes in the global order of most industries. We will only mention one of these changes here, which, from our point of view, constitutes a fundamental institutional change, namely the coexistence of two modes of appropriation. On the one hand, the classic mode is to pay for the permanent possession of a good (e.g., an iPhone, a physical book, etc.) On the other, the emerging mode is related to provision (i.e., access in the case of applications such as email, content, YouTube, or payment in cases of Spotify and Netflix). Currently, there is a certain balance between the two modes, but it is foreseeable that provision will become increasingly relevant as opposed to ownership.

As we indicated earlier, even if all groups are competing among themselves, they originally started from some particular or unique competences (i.e., DNA). In the case of Google and Yahoo, they were characterized by their search engines, Facebook and Twitter by their social networking services, Amazon and eBay by their provision of electronic commerce, while Apple, Samsung Electronics and Microsoft were originally concerned about the manufacture of consumer electronics. Finally, Nokia is a company that was initially engaged in the provision of telecommunication infrastructures. Table 2 provides some of the characteristics that identify the origins of the GAFAs.

Table 2 The origins of the GAFAs

	Year founded	Location	Founders	Current CEO	Original DNA (core business)
Google	1998	Mountain View, CA	Larry Page, Sergey Brin	Larry Page	Development of a search engine
Apple	1976	Cupertino, CA	Steve Jobs, Ronald Wayne, Steve Wozniak	Tim Cook	Designing and manufacturing consumer electronics, PCs and related software and peripheral products and networking solutions
Facebook	2004	Menlo Park, CA	Mark Zuckerberg, Eduardo Saverin, Andrew McCollum, Dustin Moskovitz, Chris Hughes	Mark Zuckerberg	Development of a social networking service
Amazon	1994	Seattle, WA	Jeff Bezos	Jeff Bezos	Electronic commerce and cloud computing
Samsung Electronics	1988	Suwon, South Korea	Lee Byung-chull	Kwon Oh-hyun	Manufacturing of electronic components and consumer electronics
Microsoft	1975	Redmond, WA	Paul Allen and Bill Gates	Satya Nadella	Development, manufacturing, licensing and supporting software products
Nokia	1871	Espoo, Finland	Fredrik Idestam, Leo Mechelin	Rajeev Suri	Telecommunications infrastructures, information technology, technology development
Twitter	2006	San Francisco, CA	Jack Dorsey, Noah Glass, Biz Stone, Evan Williams	Jack Dorsey	Online social networking to send and read short messages
eBay Inc.	1995	San Jose, CA	Pierre Omidyar	Devin Wenig	E-commerce company, providing consumer to consumer & business to consumer sales services
Yahoo	1995	Sunnyvale, CA	Jerry Yang, David Filo	Marissa Mayer	Development of a search engine

Source Own elaboration based on de Agonia et al. (2013), and Daidj (2011)

Most of these firms were created in the late 1990s or early 2000s in the US, and are at present determining the direction and intensity of global innovation dynamics to a great extent. However, European firms are underrepresented in this Internet ecosystem, Nokia being the only European group, although nowadays it is basically related to the supply of telecom access. The reality is that few European entrepreneurial firms succeed in this turbulent market. Understanding the diversification and growth strategies of the GAFAs and the high-growth start-ups, as well as their economic effects, is therefore considered crucial if Europe is to support high-growth entrepreneurship oriented toward higher innovation outputs.

Next, we provide a set of figures that give some statistics on certain structural characteristics of the groups under study for the period 2007–2014, such as the number of employees, the revenues and gross profit obtained, the value of R&D investments, the number of USPTO patents granted yearly, etc. The first indicator we will focus on is the number of employees (Fig. 2), measured in terms of Full Time Equivalence (FTE). This is a relevant indicator, not only because it allows us to see the concentration of highly skilled individuals in these organizations, but also because it provides a clear figure on their expansive and diversification strategies. It also allows us to obtain relative measures for some of the indicators we will be considering next, such as revenues, gross profit or number of patents granted. As could be expected, Samsung Electronics is certainly the largest firm, with 326,000 employees in 2014. Nokia's evolution is particularly striking, as it was the largest firm (among those considered in this chapter) in 2007 with 112,662 employees, while in 2014 this figure was halved to 61,656 people. Amazon is the second largest group with 154,100 employees in 2014, followed by Microsoft with 128,000 and Apple with 92,600. Despite Google being one of the companies that has diversified most of its activities in recent years, the number of its employees has not increased to the same extent. In fact, the major increase in the number of employees in Google can be observed between 2007 and 2011, while the figures between 2012 and 2014 basically remained constant. eBay is the next company according to size with 34,600 employees in 2014. Finally, Yahoo with 12,500 employees, Facebook with 9199 and Twitter with 3638 employees close the ranking for this particular indicator.

Next, we will provide some figures on the values observed for the revenues obtained by the GAFAs as a result of the sales of their activities, products, services, etc. As can be observed in Fig. 3, Samsung electronics dominated in this indicator during the period studied, mainly due to the broad portfolio of products manufactured, which include lithium-ion batteries, semiconductors, chips, flash memories, hard drive devices, smart phones, tablet computers, phablets, LCD and LED panels, and televisions, among others. However, according to the data for the year 2014, the sales of Apple (US$182,795 million) overtook those of Samsung electronics (US$174,883 million) for the first time. Apple showed an exponential increase in its sales, particularly following the release of the first generation of iPhones in 2007. Microsoft and Amazon rank next, with sales close to US$90 billion (in 2014 Microsoft had sales of US$86,833 million, while Amazon achieved US$88,988 million). In turn, for the year 2014 Google showed

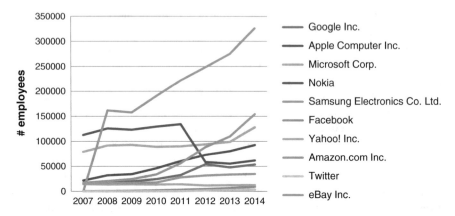

Fig. 2 Number of employees (FTE). *Source* Own elaboration

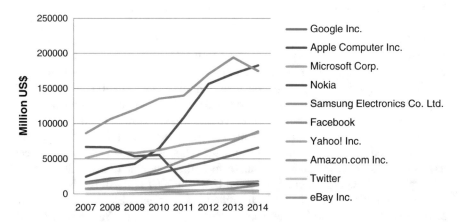

Fig. 3 Sales (revenues)—Million US$. *Source* Own elaboration

sales of US$66,001 million. Nokia's dramatic decrease in sales is also worth noting, with figures in 2007 that reached US$66,871 million, while in 2014 these only amounted to US$14,144 million. The remaining business groups showed the following figures for the year 2014: eBay—US$17,902 million, Facebook—US$12,466 million, Yahoo—US$4618 million, and Twitter—US$1403 million.

However, if we divide the values of the revenues by the number of employees, we obtain a radically different picture as compared to the previous one (Fig. 4). Apple is the company with the largest "productivity" per employee. In 2014, each employee achieved a revenue of US$1.97 million as a result of the sales of Apple goods and services. Facebook and Google are the second and third most efficient firms with US$1.35 million and US$1.23 million per employee, respectively.

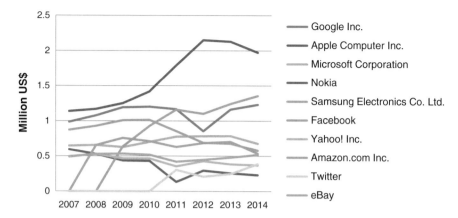

Fig. 4 Sales (revenue) per employee—Million US$. *Source* Own elaboration

The figures for gross profit (Fig. 5) show very similar patterns to those discussed previously for sales, though with some very illustrative differences for some of the business groups. While Apple (US$70,537 million) and Samsung (US$66,090 million) still lead, Microsoft is much closer to them in this indicator (US$59,755 million) than in the one concerning sales. The relative position of Google and Amazon is also worth highlighting, both of which show a very high profits (Google—US$40,688 million, Amazon—US$26,236 million) in relation to their sales. eBay (US$12,170 million) and Facebook (US$10,404 million) come next. Nokia's poorer performance can also be clearly observed in this indicator (US$6263 million). Finally, the ranking for this indicator is closed by Yahoo (US$3320 million) and Twitter (US$957 million).

However, when we analyze the figures for gross profit in relation to the number of employees in each firm (Fig. 6), the results are also rather surprising. According to this indicator, Facebook is the company that obtains the largest profits for each of their employees, with US$1.13 million/employee. In 2014 the figures for Apple and Google were rather similar (US$0.76 million), although each of these firms showed substantially different evolution as far as this indicator is concerned. In turn, Samsung, which had the second largest value for gross profit, obtained US$0.2 million per employee.

The software industry is typically regarded as R&D intensive. However, these firms do not invest much in R&D (Edquist and Zabala-Iturriagagoitia 2012). New entrepreneurial firms are not oriented toward developing long-term applications and technologies (exploration), but are instead more involved in the short-term exploitation of their competitive advantages. The latter are mostly related to the stage of the technological trajectory that the industry is currently involved in. This is a high-velocity market in which new technologies are constantly emerging and where companies must address a high degree of uncertainty. The absolute measurements of these large players' R&D investments are, however, enormous (Fig. 7). Samsung Electronics is the company with the largest investments in R&D activities

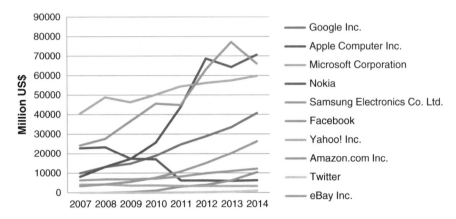

Fig. 5 Gross profit—Million US$. *Source* Own elaboration

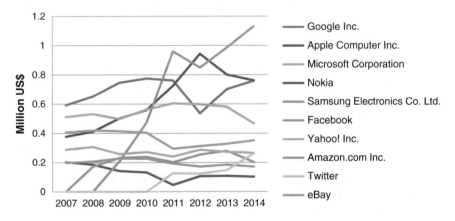

Fig. 6 Gross profit per employee—Million US$. *Source* Own elaboration

with US$12,200 million, closely followed by Microsoft with US$11,381 million. Google and Amazon also show similar patterns over time with US$9832 million and US$9275 million in 2014, respectively. However, despite the similarities that Apple often shares with Samsung, it only invests half as much, registering US$6041 million in 2014. Nokia's evolution for this particular indicator is also very similar to that already noted in relation to the number of employees. In 2007, Nokia was also the player that invested the most in R&D (US$7657 million), while in 2014 its investments amounted to US$2769 million. Finally, we would also like to stress Facebook's evolution. While in 2009, Facebook only invested US$84 million, in 2014 this figure rose to US$2644 million, after having remained constant between 2012 and 2013.

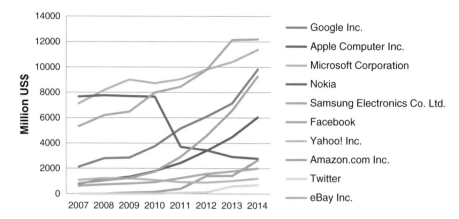

Fig. 7 R&D investments—Million US$. *Source* Own elaboration

However, as is also the case with R&D investments at the country level, we need to provide a relative figure, usually compared to the level of the gross domestic product to obtain conclusive insights about its relevance in the economy. Given that we cannot talk about gross domestic products in the case of firms, we will provide this relative figure as a ratio of the GAFAs' R&D investments over sales. As was the case with the previous relative figures, here too the patterns noted for each of the business groups also show a radical change. In this case, it is Yahoo where R&D investments played the most central role in the firm's strategy in 2014 with 26.14 % of the sales being devoted to R&D activities, although these figures have undergone an important change since 2012, the year that Marissa Mayer was appointed as the CEO of the firm. Facebook is ranked second with 21.21 %. The relative position of Nokia is also worth highlighting here. As we discussed earlier, Nokia has undergone a dramatic reconversion, particularly since 2007. However, as can be observed in Fig. 8, and against all expectations, Nokia continuously increased its share of R&D investments, in particular between 2010 and 2011, years when the company was considered to be bankrupt, out of the smartphone race and when the deal with Microsoft was about to be signed. Google and Microsoft have followed parallel paths, particularly since 2011, with shares close to 15 %. The same can also be said regarding eBay and Amazon, which converged in 2014 around 11 % after embarking on quite different paths.

As we discussed earlier, one of the elements that best characterizes the GAFAs is patents, particularly those granted at the USPTO due to the possibility of patenting software. Figure 9 provides evidence of the number of patents granted per year at the USPTO to the GAFAs between 2007 and 2014. Here again, when we observe the gross value for the patents granted yearly, it is Samsung Electronics which undoubtedly leads (5794 patents in 2014), outperforming the values of Microsoft (3161 patents), Google (2659 patents) or Apple (2195 patents). Nokia's performance has remained somewhat constant over these last years, with values close to 900 patents, while Amazon has significantly increased the number of patents granted on a yearly basis, particularly since 2010, reaching 751 patents in 2014.

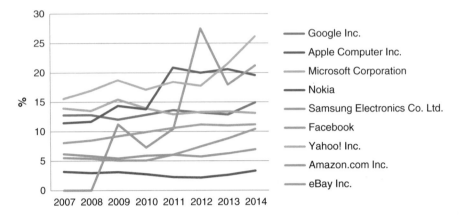

Fig. 8 Share of R&D investments on sales. *Source* Own elaboration

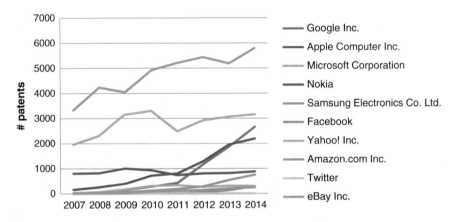

Fig. 9 Number of USPTO patents granted. *Source* Own elaboration

In order to provide evidence of the relevance that patents have in these business groups' strategies, we next provide relative figures for the number of USPTO patents granted per billion US$ invested in R&D (Fig. 10) and the number of USPTO patents granted per thousand employees (Fig. 11). With these measurements we intend to provide a relative view of the efficiency of R&D investments in relation to patents on the one hand, and on the productivity of employees on the other. In the former case, it can be observed that in 2014 Samsung Electronics was the company with the highest efficiency in its R&D processes, as it obtained 475 patents per billion US$ invested in R&D. However, the evolution is quite negative when compared to the firm's performance in 2008, where it achieved 684 patents per billion US$. The opposite is the case for Apple, which doubled its efficiency in the period under analysis, going from 200 to 363 patents/billion. Similarly, Nokia tripled the efficiency of its R&D investments. While in 2007, 105 patents were obtained for each billion US$ invested in R&D, in 2014 this measure rose to 317, overtaking Microsoft, Yahoo, and Google.

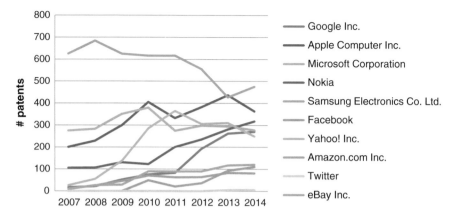

Fig. 10 USPTO patents granted per billion US$ invested on R&D *Source* Own elaboration

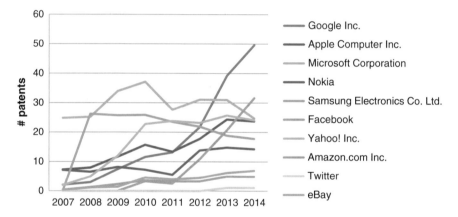

Fig. 11 USPTO patents granted per 1000 employees. *Source* Own elaboration

As for the productivity of the labor force in relation to patents, it was Google that led in 2014 with 50 patents granted per thousand employees. Facebook was the second company in this respect with 31 patents/thousand employees, although the significant gains in productivity since 2011 must be noted. Microsoft, Yahoo, and Apple were granted 25 patents/1000 employees. However, while Microsoft returned to the same productivity registered in the year 2007, Apple, and particularly Yahoo, more than tripled theirs. In this case, due to the large size of the company in terms of employees, the relative performance of Samsung Electronics was quite modest (18 patents/thousand employees in 2014) as compared to its counterparts. Nokia's good performance was also remarkable; in 2014 it doubled its figures from 2007.

The final indicator we focus on here is these firms' market capitalization (Fig. 12). With this figure we aim to determine public opinion regarding the net worth of these companies. In this case, it is Apple that leads by an astonishing difference. In 2012 Apple's shares had the highest value on the stock market with an average

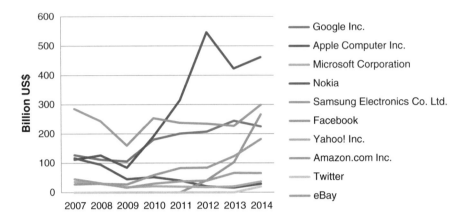

Fig. 12 Market capitalization (billion US$). *Source* Own elaboration

capitalization of US$546 billion. Samsung is the second most valuable company according to this indicator with a market capitalization of nearly US$300 billion, while Google ranks fourth with US$225 billion. The most significant trends are those exemplified by Facebook and Amazon. While the former has become one of the major players on the stock market since its initial public offer in 2011, Amazon showed an interesting trend resulting in a market value of US$182 billion in 2014.

3.1 Illustrating the Business Ecosystems of the GAFAs

As we discussed in Sect. 2.1, the GAFAs can be regarded as business ecosystems rather than mere business groups, due to the large diversity of players embedded in their environments, processes, value chains, goods, and/or services. Next, we will introduce graphic illustrations of the business ecosystems of some of the firms we are interested in.[16] Although they may not represent the current state of their respective ecosystems, due to their high variability and patterns of change, we consider them useful for obtaining an initial picture of their original activities, products and processes, which will help us to better understand their epigenetic moves in the next sections. For example, Google and Amazon are not heavily involved in hardware, although Google is indirectly involved in hardware through its Android mobile operating system. Amazon rules when it comes to retailing, particularly regarding books. Apple and Microsoft are relatively weaker in shopping, but Microsoft is strong in gaming. Apple is hardware-centric. Contents and applications available on iTunes produce revenue and profit, but are all oriented to providing services for the hardware. In a sense, Apple is a counterpart of Amazon,

---

[16]No graphical evidence could be found for Yahoo, Samsung Electronics, Nokia and eBay's business ecosystems.

as the latter's business is the sale of content. Thus, Apple and Amazon have a business model in which the key is the sale of goods or content. Facebook and Google mainly obtain their income from advertising. Facebook and Google are data-centric. However, one way or another, they are all locked in a death match (de Agonia et al. 2013), particularly concerning dimensions such as books, gaming, music, video, and online shopping, although the boundaries between each of them are hard to define.

As comprehensively evidenced by de Agonia et al. (2013), "during its first decade, Google made massive headway in the field of Internet search, an area that had remained relatively stagnant until its arrival" (de Agonia et al. 2013), and which had been dominated by the search engines of Altavista and Yahoo. Google's first steps in the media world were taken in the fall of 2006, when the firm announced it was acquiring YouTube. The company released its plans for the Android mobile operating system in 2007, a year later launching the first generation of phones that incorporated Android software. Google also unveiled a web-centric operating system called Chrome OS in 2009, a television platform called Google TV in 2010 and services for online movie and music streaming in 2011 (ibid). However, other platforms such as Google Ads, Google+, Google Play Store, Google Books, Google Maps, Google Drive or Google Wallet are also part of Google's ecosystem (Fig. 13). With each move, the simple search box moved further, although search itself remained at the core of the company's business.

The Google Play Store has robust sections devoted to book, magazine, and music related content (i.e., Google Play Music All Access).[17] As of July 2015, the Google Play Store included more than 1,600,000 apps (Statista 2015). In 2014, the top five categories of apps by revenue at the Google Play Store were: gaming, communication, social, tools, and travel (App Annie 2015). Google also has a separate service named Google Books, which makes it possible to search for content within actual books and magazines.[18] However, one of the pivotal elements of the current Google ecosystem is the Android OS. Since it is an open-source platform, anyone (be they a large established organization, an entrepreneur, a developer, etc.) can use the software in any potential device (i.e., already existing or not) and modify it in any way. As a result, despite Google not being explicitly present in gaming (yet), a number of manufacturers offer Android-based devices made specifically for entertainment purposes (e.g., Sony's Xperia Play phone, Nvidia). In this regard, the Google Play Store offers a large collection of apps, some developed by Google, and most of them developed by the millions of developers worldwide that use the Android platform to develop games.

---

[17]As explained by de Agonia et al. (2013), "Google Play grew out of the former Android Market, which was essentially an app store for Android-based phones and tablets. As the market expanded to include more types of content, Google wanted a name that'd fit the broader focus and emphasize the fact that the store wasn't limited just to Android users".

[18]As de Agonia et al. (2013) note, "Google Books has been the subject of much controversy within the publishing world because of the complex rights issues related to Google's scanning of older print editions".

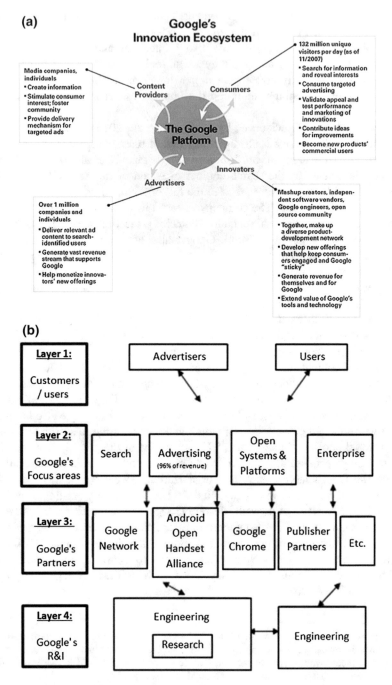

Fig. 13 a Google's ecosystem (2008). *Source* Iyer and Davenport (2008). **b** The core of Google's Global Innovation Ecosystem (2014). *Source* Fransman (2014: 24)

In 2009, Google announced the launch of the Chrome OS, a platform in which applications and user data are stored in the cloud. The most common products that use the Chrome OS are Chromebooks, which run on web-based applications. YouTube has become the key source for searching web-based video, which ranges from entertainment videos produced by big-name studios (e.g., movie producers) to homemade clips from individuals across the world, many of whom use the platform to become worldwide "stars" (e.g., Smosh, The Fine Bros, Ryan Higa). In parallel, Google Play offers a variety of possibilities for renting and/or purchasing movies and TV shows (e.g., full seasons or specific episodes).

The cross-platform nature of Google's ecosystem allows the company and its users to gain access to any content from any connected devices at any place and any time, a competitive advantage that many other GAFAs are also putting into practice.

In the case of Apple, the devices (i.e., all mobile hardware) they manufacture are the key to understanding its "digital hub" ecosystem (de Agonia et al. 2013). In 1984, Apple's Macintosh was the first mass market computer that incorporated a graphical user interface and a mouse. After that, a flurry of technology initiatives (e.g., iMovie, iTunes, iPhoto) have emphasized Apple's intentions to bring computerized media to users (ibid). In this regard, Apple's ecosystem cannot be understood without its first digital player, the iPod (2001), that shortly afterwards was accompanied by the release of iTunes (2003), which later entered the current App store. In 2007, with the announcement of the iPhone, Apple created a disruption that radically changed the way telephony (i.e., communication, mobility, and social relations) was understood. This breakthrough was followed by the release of the Apple TV and the iPad. Finally, the last of Apple's products to date has been the iWatch (2015) after a long period of development and maturation.

Each of these devices could be understood as platforms, which provide access to a large variety of online content and services (i.e., music, video, TV shows, games, mobility, banking, health, sports, retailing, software, etc.). It is in these platforms that the actors in the Apple ecosystem (e.g., developers, telecom providers, other large firms Apple cooperates with such as Nike) can introduce their new ideas, projects, services, developments and thereby "co-evolve" and "co-shape" the ecosystem they are part of (Fig. 14).

Apple entered the world of digital books in 2010, when the iBooks app was introduced as a gateway for the iBookstore (de Agonia et al. 2013). However, within the App Store there are thousands of apps listed under the books category, which also provide support for competing services, like Amazon's Kindle, and many other e-book readers.[19]

At the moment of writing this chapter (August 2015), Apple's app store included more than 1,500,000 apps (Statista 2015). In 2014, the top 5 iOS App Store categories by revenue were: gaming, social networking, music, education

---

[19]As discussed by de Agonia et al. (2013), back in 2012 Apple was having "some trouble with the government for alleged antitrust actions having to do with e-books… the U.S. Justice Department charged that Apple spearheaded a scheme with book publishers to keep the prices of e-books artificially high; on July 10, Apple was found guilty of violating antitrust laws".

and entertainment (App Anie 2015). The number of apps available at the App Store and the Google Store and the repercussion that both OS (iOS vs. Android) have on the smartphone industry lead us to consider that there is battle between the Apple and Google app stores (OECD 2013).

However, in spite of the clear success of mobile gaming, the same cannot be said about Apple's desktop-based gaming, where Apple does not have a console to compete against Microsoft's Xbox, or Sony's Play Station (de Agonia et al. 2013). Similarly, Apple does not yet have a central place where users can buy whatever they may be interested in, as is the case with Amazon and Google. Finally, iCloud is the platform used by Apple to allow for online storage of content (e.g., music, video, books, podcasts, apps). In addition, it also allows all Apple devices to be synchronized so that changes made on one device are automatically transferred to all the other devices belonging to the same user.

These strong innovation capabilities have a massive influence not only on Apple's evolution, but also on its competitors' strategies, which are in some way obliged to follow Apple's logic. However, as we will see, the evolution of these global business ecosystems also has a profound influence on, and causes severe consequences for, many other industries (Evans et al. 2008).

Figure 15 provides an illustration of Facebook's ecosystem, where each icon represents a specific area or action that users can take on Facebook, ranging from apps to photos and events (Trewe 2011). In this way, Trewe provides evidence of the variety of actions users take on their social platforms. However, it should be noted that this version of the Facebook ecosystem is not complete as the network is intricate and changing rapidly.

Facebook began as an online social networking service in 2006, mainly accessible by computers, although it was soon adapted to the changing characteristics of mobile devices as well. As of August 2015, 1490 million accounts were available on Facebook, while Google+ only reached 300 million users and Twitter 316 million (Statista 2015). In 2012, Facebook announced App Center, a store selling applications that operate via the site. Facebook's ecosystem encompasses users, advertisers, other social networks, developers, suppliers, and operating systems (Bonde 2013). However, advertising is pivotal in Facebook's business ecosystem. In fact, both Facebook and Google compete in this advertising market, the two of them accounting for the most significant share of online advertising space. In the fourth quarter of the year 2012, more than US$1.3 billion of Facebook's revenues came from ads.

Facebook is capable of showing the right commercial and noncommercial content to the right person, at the right time and location (ibid: 31). Facebook is capable of adapting to the content distribution method that best suits users and advertisers. Thus, Facebook's data analytic capabilities are at the core of the company. According to Bonde, in 2013 there were more than one million active advertisers on Facebook. Advertising on Facebook is valuable not only for direct sales, but also for creating brand effects and for data creation (Bonde 2013). In this regard, the largest advertisers in social media are financial services, travel and leisure, consumer packaged goods, information, computing, electronics, and retail (ibid).

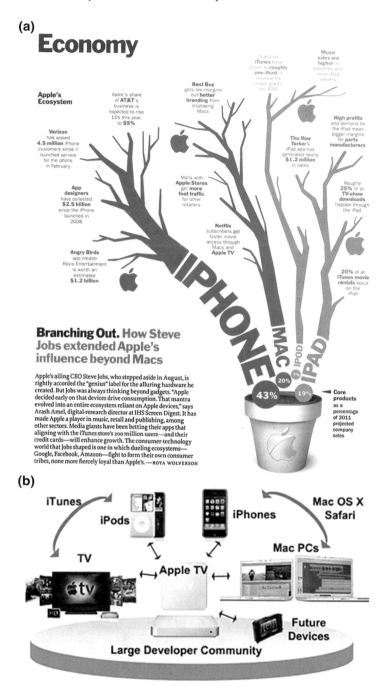

Fig. 14 **a** Apple's ecosystem (2011). *Source* TIME, September 12, 2011. Available: http://obam apacman.com/2011/09/time-magazine-apple-ecosystem-infographic/. **b** Apple's innovation ecosystem (2014). *Source* Nielson (2014)

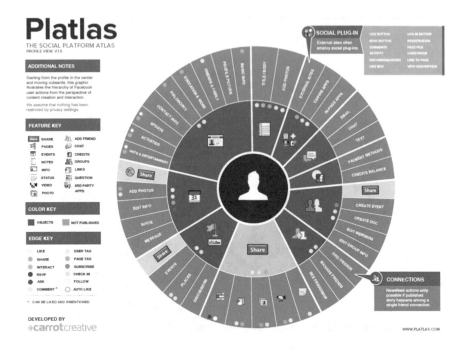

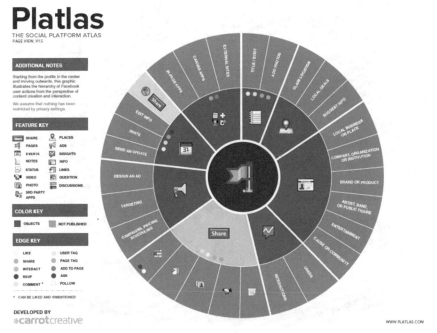

Fig. 15 Facebook's ecosystem (2011). *Source* Trewe (2011)

As discussed by Cocotas (2013: 6), social advertising runs on a freemium model, so that users can join for free and then only pay for premium services. In this regard, a consensus seems to be forming around in-stream advertising as the most promising social advertising format (ibid).

Facebook offers several ways to advertise in its platform: brand pages, display ads, sponsored stories, promoted posts, page post ads, mobile app install ads, and log-out screen ads. In addition, it has introduced a series of tools that brands can use to obtain data on their target customers' uses and habits, such as custom audiences, partner categories, cost-per-action or the Facebook exchange system (see Cocotas 2013: 16).

Entertainment media is part of Amazon's basic DNA (de Agonia et al. 2013). Amazon was founded as a book-selling site, to then branch out into other products such as CDs, DVDs, videogames, electronics, apparel, baby products, consumer electronics, beauty products, gourmet food, groceries, health and personal-care items, industrial and scientific supplies, kitchen items, jewelry and watches, lawn and garden items, musical instruments, sporting goods, tools, automotive items, and toys and games. M&As are one of the means that Amazon has used in its diversification strategy. Today, Amazon is a retail behemoth, but media remain its core business. For example, the Kindle, one of its key products, is designed to make it easier to buy not only books, which it was originally designed for, but also movies, TV shows or music through the Amazon platform (de Agonia et al. 2013). So, it can be said that its media ecosystem is powerful, well-integrated, and possibly the largest in the world (ibid).

Amazon is more than just a retailer of physical products (e.g., books) or downloadable ones (e.g., games). Its ecosystem (Fig. 16) also offers several imprints, tools to help musicians create their own music, and it is producing its own TV shows as a way to attract people to its streaming video service (de Agonia et al. 2013). Since 2000, it has also included Amazon Marketplace, a platform that lets customers sell used products alongside new items. In 2011, Amazon announced its entry into the tablet computer market by introducing the Kindle Fire. Besides, in 2014 Amazon announced its Amazon Fire TV and the Fire Phone, thus entering the TV and smartphone markets, respectively, essentially following in the steps of Apple.

In 2011, Amazon launched the Amazon Appstore, which in August 2015 included more than 400,000 apps (Statista 2015). In 2013, the company announced its Mobile Ads API for developers, which can be used on apps distributed on any Android platform as long as the app is also available in Amazon's Appstore. However, gaming is not one of the key streams of activity Amazon aims to pursue with its ecosystem, in contrast to Apple, Google, or Microsoft.

To date, besides selling, which still remains at the core of Amazon's ecosystem, one of its pivotal elements is cloud computing, a market that the firm first entered in 2002 with the launch of Amazon Web Services (see Case Publisher 2008). Since then, Amazon's cloud computing platform includes services such as the Elastic Compute Cloud or the Simple Storage Service, which are used by organizations worldwide in need of heavy computing resources such as banks.

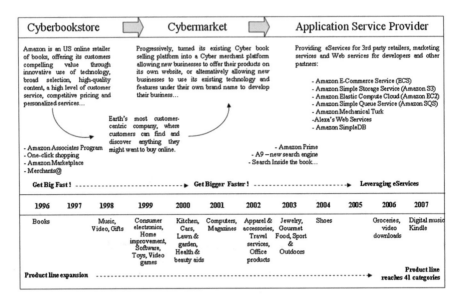

Fig. 16 The evolution of Amazon's business model. *Source* Isckia and Lescop (2009: 45)

Together with Apple (1976) and Nokia (1871), Microsoft is the oldest of all the GAFAs we consider in this chapter, as its creation dates back to 1975. The original purpose of the company was far from being entertainment-oriented, but instead was focused on "products such as a BASIC programming language interpreter, and then (with a contract in late 1980 with IBM) on operating systems like PC-DOS" (de Agonia et al. 2013). Microsoft, which was regarded by many as the evil to be fought against (which gave rise to the entire open-source movement, originally with Linux),[20] has had both major successes and failures. For example, "the Xbox 360 video game platform and community have become a rousing success, along with the Halo and Age of Empires series of games. Microsoft seems to be holding its own with its Windows Phone mobile devices, but there are a lot of people wondering whether its Surface tablets are innovative or a mistake. Still, Microsoft's media system is a work in progress, being very strong in some areas such as gaming, and very weak in others, such as shopping and books" (ibid).

Similar to the App Store and Google play, in 2004 Windows launched the Windows Marketplace, a platform for the delivery of software electronically rather than physically as was previously the case with the Windows OS. In 2012, it was replaced by the Windows Store, an application store for Microsoft Windows and other types of apps. As of August 2015, there were 340,000 apps available in the Windows phone store (Statista 2015). Besides the apps, Microsoft has also put considerable effort into developing cloud computing capabilities in-house. In this regard, the company released the Microsoft Azure platform in 2010.

---

[20]Also note Google's corporate moto: "Don't be evil".

Microsoft's gaming platform, anchored by the Xbox 360 console, is certainly the most comprehensive of all the GAFAs. Microsoft has put considerable effort into making its Xbox 360 the center of its business ecosystem, integrating it with other Windows-operating devices (Fig. 17). This was the goal sought with the launch of the Xbox One in 2013, which not only allowed for gaming, but also for interacting with other media content such as music or video. From being a company that mainly ran on licensing proprietary OS, Windows has become a big game platform (de Agonia et al. 2013).

Microsoft has also included a powerful search engine in its ecosystem: Bing. Besides carrying out Internet searches, it also includes the Bing's Shopping feature and the Wallet technology, the latter resembling the characteristics previously discussed on Google or Amazon's ecosystems. The company has also engaged in communication activities through Skype, its voice and video communications tool, which could at some point play an important role in Microsoft's entertainment/gameplay platform (de Agonia et al. 2013). Finally, the acquisition of Nokia's Devices and Services business unit by Microsoft in 2014 has enabled the firm to become an important player in the smartphone industry, mainly through the Windows Phone and Surface tablets (Risku 2012).[21]

Twitter was founded as an online social networking service and micro-blogging platform that enables users to send and read short 140-character messages called "tweets." One of the central elements for Twitter's success is its ability to track tweets in real time. In this sense, tracking the ten most-talked-about topics at a given moment has been labeled as "trending topics," making it possible to follow such topics in different geographical zones (i.e., worldwide, a particular country).

Registered users can read and post tweets, but unregistered users can only read them. Users can access Twitter through the website interface, SMS or the apps for most mobile devices available on the market. As Bmimatters (2012) states, "content and media companies are using Twitter to drive traffic to their websites. It is being used by e-commerce and local businesses for deal promotions. Some businesses are using it as a customer service channel; while some are using it to increase their brand awareness and monitor their brand perception. Some non-profits are using Twitter as a fund-raising channel as well" (Fig. 18).

In August 2015, Twitter had 316 million users (Statista 2015), thus becoming an attractive destination for advertisers. Primarily, the efforts of Twitter are oriented to establishing partnerships with search vendors, device vendors, media and telecom providers (Bmimatters 2012). In fact, Twitter's business model also runs on the advertising market.[22] Twitter's advertising efforts are analogous to Facebook's efforts, and it could be said that, to a certain extent, Twitter's ad products mirror Facebook's. Twitter has three primary ad formats: promoted trends, promoted accounts, promoted tweets and keyword targeting (see Cocotas 2013).

---

[21]When the deal between Microsoft and Nokia was reached (US$7.2 billion acquisition), Microsoft acquired a patent portfolio of up to 8500 design patents, but not the many other thousands of the Finnish company's utility patents, which were licensed to Microsoft for 10 years.

[22]For more details on the economics behind Twitter's business model see Levy (2015).

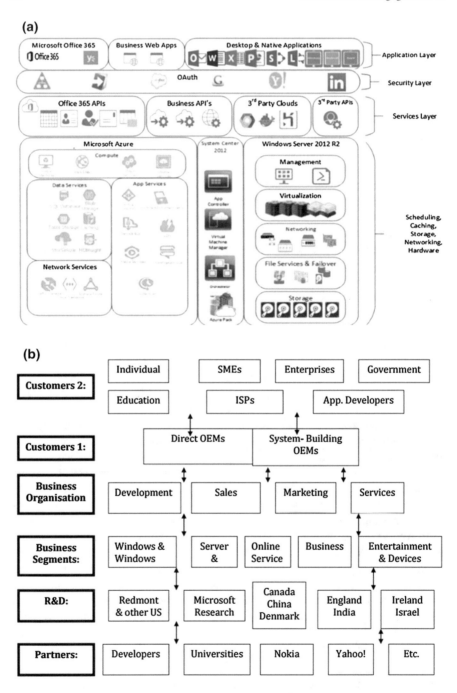

Fig. 17 a Microsoft's business ecosystem. *Source* Skelly (2014). **b** The core of Microsoft's Global Innovation Ecosystem. *Source* Fransman (2014: 27)

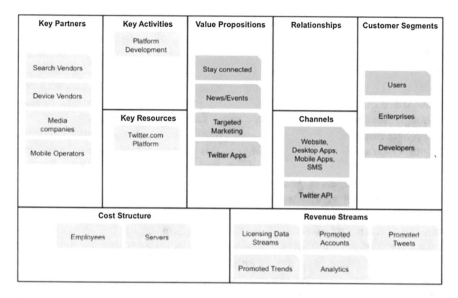

Fig. 18 Twitter's business ecosystem. *Source* Bmimatters (2012)

According to Cocotas (2013: 31), video ads would be a natural expansion for Twitter. This is an area where Twitter is already being challenged by Meerkat, so it seems plausible Twitter will engage in this stream of activity in the short-term in order to defend itself from this increasing competition.

Twitter has also built an app ecosystem, offering APIs that help developers build third party apps. As the Twitter official blog indicated on July the 11th 2011, "As an ecosystem, we've just crossed one million registered applications, built by more than 750,000 developers around the world… A new app is registered every 1.5s, fueling a spike in ecosystem growth in the areas of analytics, curation and publisher tools."

4 Epigenetic Dynamics

Every day we hear about a new business activity in one (if not more) of the GAFAs, which was already carried out by another group. As discussed, we believe they will gradually reach the point where they are all doing the same business activities, although with varying degrees of penetration (and success), and always keeping to a main activity or function (i.e., original DNA). However, this business convergence is at the same time leading them to diversify to new sources of income and risks. They are increasingly being keen to enter fields that do not depend on their original activities. When one group starts up a new business venture or starts exploring a new direction (i.e., what in the context of this chapter we refer to as dominant vector), the others imitate and follow (Daidj 2011).

One example of this convergence would be the availability of social networks. Part of the business groups we are aiming at studying are present in the field of social networks. Facebook, because it is the group's DNA. Google, because it improves and complements user data in searches. Amazon, by using sales data mining and recommender systems. Kindle also enables users to contact with people who read the same book, and Apple integrated Twitter into its new devices. Although Amazon and Apple cannot equal Facebook's social power, they could set up agreements with the firm or, even more easily, with some competitor like Twitter. Other examples (among many) of this convergence could also be the provision of videophone services, or the increasing proliferation of eBooks some years ago and of tablets in more recent times.

This is the goal of this section. To illustrate the epigenetic dynamics of the GAFAs over the past years, providing evidence of the extent to which the moves by some of them are also followed by the others. This section thus responds to the second stage of the EED approach, which focuses on analysing the dynamics in response to influences from the environment, as a result of which "genetic disorders" may be created. These deliberate changes occur abruptly rather than gradually. In other words, epigenetic dynamics follow an economic rationality. They do not happen by chance as it might be deduced from the Darwinism that has dominated evolutionary thought. We would like to stress here the need to constantly follow up these moves so as to get a comprehensive and updated view of the strategy and dynamics of each of these firms. This is particularly central in the analysis of the digital ecosystem. Thus, it might be the case that the dynamics we are analysing in this section become obsolete in a short period of time, and that the dominant vectors in which the GAFAs (or those which still remain solid in the marketplace) are putting their efforts into in say 3–5 years time are radically different from those discussed here. This is why the book is trying to balance the conceptual and empirical parts in the book, contributing with an analytical framework that allows interpreting these dynamics, no matter which these are, in the light of the literature on evolutionary economics.

Entry of these big business groups in other corporations' fields may though be seen as a defensive rather than offensive move. On the one hand, they are looking to diversify, because as Manjoo (2011) pointed out, "You never know what's going to hit big in tech. So if you can, why wouldn't you try everything?" Furthermore, Google may be concerned because some analysts like Vincent Cerf, one of the so-called fathers of the Internet pointed out that the firm's essence, the search engine, may be overtaken by another, just as Altavista, the best engine in its time, was displaced by Yahoo, which was then surpassed by Google. As we have indicated in Sect. 2 in this chapter, due to the increasing proliferation of applications in the Internet ecosystem (i.e., particularly concerning mobile) search engines could suffer a major disruption in the short run.

Iansiti and Richards (2006) make an analogy between competition and evolution in the sense that some animal species "run a race" of adaptation in their evolution. This enables them to defend themselves from their predators to avoid their extinction as a species. From an evolutionary perspective, competition for the

GAFAs implies permanently resizing and readapting to maintain an identity and a place on the market, which requires aggressive strategies (Mortara and Minshall 2011). Some of the characteristics of these epigenetic dynamics are included in Block B of Table 3. These dynamics are produced, among others, as a response to the characteristics of the environment mentioned which these firms operate, namely, the digital ecosystem: Intense increase in inter-group competition.

Innovation becomes an asset and a key activity. This has been true for a long time now. What is new is the difficulty companies experience keeping up with the extremely fast pace of innovation for long periods of time. Companies like Nokia or RIM (Blackberry) that were leaders just a short time ago have been displaced by others such as Samsung, Apple, or Google, and their respective business eco-systems, which have drastically changed the mobile telephone landscape and the overall Internet panorama. Nokia has not been overtaken by another mobile tel-ephone company but by an ecosystem in which the device is merely a part of it (Gruman 2012; Thomas and Autio 2012). Accordingly, as previously discussed, the companies we are interested in need to define and deploy comprehensive, intense, and complex innovation strategies (Iyer and Davenport 2008). Even so, they are not enough to manage innovation of the entire ecosystem, especially as the business groups move away from the main function of the ecosystem (i.e., the perimeter of innovation).

Table 3 An epigenetic understanding of the economic impact of big internet business ecosys-tems' dynamics

A. Analysis of the environment	B. Identification of epigenetic dynamics
Intense increase in inter-group competition	Massive acquisition of small firms and/or their intellectual property (i.e. patent portfolio) to block potential structural changes and to defend from competition
Exponential growth of the markets and users in other (related) business areas	Aggressive acquisition strategies to sustain profit growth
Expansion in the number of applications and their content	Presence in global markets and gain access to new technologies and innovations
Fast multivectorial technological change and planned obsolescence	Asymmetric negotiations between large business groups, application developers and content providers
Modularity in the behavior of business ecosystems	Entry of large business groups in activities not related to their original purpose (DNA)
Exponential increase in advertising as a share of turnover	High entry barriers posed by large incumbents
Exponential increase in the patent portfolio	Risk-averse industrial strategies played by large companies
Dominance by large companies that are increas-ingly being delocalized	Use of the financial strength as the main pro-tective industrial instrument
Industry, market and institutional structures	

Source Own elaboration based on Gómez-Uranga et al. (2014)

Mobile telephones are essential to all the groups we are interested in analysing, due to the possibilities of future expansion. For Facebook and Google in the field of advertising; for Apple, because it is the firm's key business, needed to maintain its leadership in a wide range of innovative products and discover new sources of income (amplified iTunes, iCloud, etc.). In fact, big changes in business organizations' environments have been observed with the arrival of the new 4G technology as it marks a qualitative break for mobile telephones. The promising future of mobile phones prompted Amazon to consider launching Kindle, which, of course, made it easier to sell their goods to users who purchased it. It is also a support device that could be used for other activities, such as advertising. Google and Amazon had also launched their own replicas of iTunes and more recently of the iCloud (Shearman 2012). Television may also be a dominant growth vector in the future, although at current times the main dominant vector is that related to cloud computing.

When observing the activities of big Internet industry groups it can be noticed that their development does not conform to the natural/gradual processes that evolutionary economics would predict (see Chapter "Introducing an Epigenetic Approach for the Study of Internet Industry Groups"). These large groups were originally distinctive for know-how that resulted in some "initial business routines" as well as certain products that fit into their "original activity." However, these groups suddenly and rather abruptly took on new routines and entered fields that initially had nothing to do with their businesses. Table 4 illustrates the convergence observed in the direction and intensity of the dynamics of some of the leading Internet business groups between 2012 and 2013.

More recently though, these dynamics have even become more disruptive, as the business groups have started to penetrate other domains which are not that close to the software industry as Table 4 illustrated. As we can observe in Table 5, some examples of these radical and abrupt economic dynamics are represented by the increasing moves toward health and genetics, self-driving cars, augmented reality, higher education, or finance and banking to mention a few. As discussed earlier, the presence of these large business groups in these sectors cannot be explained or expected according to their original activities or DNAs, but rather from the pressure introduced by competing firms and the environment in which these operate.

Internet firms are obliged to follow these dominant vectors because they cannot leave areas or segments outside their organizational routines if they are to survive in the marketplace. Even if that behavior leads them to have an inefficient behavior, they need to penetrate other (often unrelated) sectors of activity because in their absence they would be relegated from other areas where rival groups dominate.

The previous diversification strategies have been motivated by the convergence context (Daidj 2011). Apple has always adopted an "original" Internet pay model (subscription model for content in iTunes) compared to its competitors. Between 2011 and 2012, Apple released the iPhone 4S, iPhone 5, Siri, cloud-sourced data services with iCloud, the third and fourth generation iPads, the iPad MiniIn, iBooks Textbooks for iOS and iBook Author for Mac OS X, a third-generation MacBook Pro and new iMac and Mac Mini computers. In 2013, Apple got deeply involved in the development of augmented reality systems, following the steps by

Table 4 The companies' activities in different fields (market shares and ranking by activities or products) between 2012 and 2013

Business Groups	Hardware		Software	Contents	
	Mobile /Smart phones[a]	Tablets	Op. System—also for smart phones	Search engines	Social networks
GOOGLE	Nexus	Nexus	Bada 5th	**	65 m
			0.7 %	1st	
	Smart Phone		Android 1st	88.8 %	
			74.4 %		
APPLE	**		IOS		
	3rd	1st	2nd		
	9.0 %	39.6 %	18.2 %		
MICROSOFT	Expected Smart Phone with Nokia	Surface,os:rt	**	Bing	
		5th	4th	2nd	
		1.8 %	2.9 %	4.2 %	
FACEBOOK	Expected Smart Phone with HTC			4th in the US	**
					1st
					750 m
					72.4 %
AMAZON	Expected	4th			
		3.7 %			
SAMSUNG	1st	2nd	Android		
	23.6 %	17.9 %	1st		
			72.4 %		
NOKIA	**	Expected	Symbian		
	2nd		6th		
	14.8 %		0.6 %		
HTC	8th				
	2.3 %				
MOTOROLA	9th				
	2.1 %				

Convergence in the direction and intensity across Internet Business Groups
[a]The figures included in the first column refer to the Worldwide Mobile Phone Sales to End Users by Vendor in the first quarter of 2013. For the Worldwide Smartphone Sales to End Users by Vendor see Gartner (2013)
**Activity/Routine within the DNA of the business group. The first "innovative" products that best identify with the Internet industry groups' DNA would be those that form the essence (i.e., products that each company has been distinctive for since its beginnings)
Source Own elaboration from Gartner (2013), Alexa Global traffic Rank (2013), IDC (2013) and Gómez-Uranga et al. (2014)

the other GAFAs, particularly Google and Facebook, and more recently wearables. More recently, the company is getting into the manufacture of smart electric cars and the production of green energy, and getting serious about video games through the new developments carried out in the new Apple TV.

Table 5 Economic dynamics in some of the leading Internet business groups in 2014

Business groups	Sectors						
GOOGLE	The cloud	Wearables, health and genetics	Self-driving car	Connected homes and societies	Digital ads	Augmented reality	Higher education (udacity), Finance and banking
APPLE	The cloud	Wearables and health	Third-party Digital content (music, movies, games, podcasts)	Apple Smart TV			
MICROSOFT	The cloud		Gaming	Social networks		Augmented reality	
FACEBOOK	The cloud	Wearables (Fitness)	Private social networks	Connected homes and societies	Digital ads	Press and media	
AMAZON	The cloud	Online payment systems	3D printing	Amazon TV and Amazon Studios	Digital ads		
SAMSUNG		Wearables and health	Samsung Bikes	Connected homes Smart TV	Microchips for smartphones	Augmented reality	Wi-Fi standards
TWITTER			Finance and banking	Video communication			

For a more detailed analysis of the most recent dynamics included in Table 5, see Chapter "GAFAnomy (Google, Amazon, Facebook and Apple): The Big Four and the b-ecosystem"
Source Own elaboration

Google has adopted a diversification strategy of nonrelated media products and services in such a way that hardware and software have become more inextricably linked at the company (Finkle 2011). Google has also employed the Web Search technology into other search services, including Image Search, Google News, the price comparison site Google Product Search, the interactive Usenet archive Google Groups, Google Maps, Picasa, Orkut, Youtube, Google books, Google Scholar, Google Patent search, Google Docs, Google Chrome, and Chromebook. In 2013, the company announced the launch of Calico, a firm focused on the challenge of aging and associated diseases in a clear move toward the health sector and the pharmaceutical industry (i.e., Google X project, Google Fit). A first step in this direction has been the development of wearables by Google, following the steps undertaken by Apple and Samsung among others (see also Peeble).[23] In 2014 the firm announced the acquisition of Nest in order to more into the Internet of Things and the connected house. The more recent moves of the firm into the higher education (i.e., university education) and academic research segment should not be overlooked, due to the follower effects it can create in the other GAFAs. Partly derived from the previous epigenetic dynamics during the past years, in 2015, Google announced plans to reorganize its various interests in a holding company called Alphabet Inc. give operating divisions more leeway in making their own decisions and keep the businesses more nimble (Dougherty 2015). From our EED approach, this corporate reorganization can be regarded as one of the most significant epigenetic dynamics that may occur. This reorganization in holdings undertaken by Google may generate some potential consequences such as: the existence of a magma business with different yields and expectations; a difficulty to identify and separate each product complexity and therefore to calculate the actual value of each business by investors; regulatory problems. One of the dominant vectors that may guide the future of Google, together with Facebook, is the development of systems that allow satellite Internet connection (i.e., Space X and Project Loon). Another is Kobalt, with the one the company is intending to disrupt the music industry. The other, namely, the development of driverless cars is also being fostered by the investments Google has carried out in building artificial intelligence systems (i.e., TensorFlow) that can learn from video games.

As early as 2002, Microsoft decided to be the first mover in the promising online game sector, developing a specific OS for the Xbox 360. This led the firm to the development of Kinect. It is expected that Microsoft might use Kinect to penetrate in the TV industry as Amazon and Google are increasingly doing. In this sense, in 2014 Microsoft acquired the video game development company Mojang (i.e., Minecraft) for $2.5 billion. Following the release of Windows Phone, Microsoft underwent a gradual rebranding of its product range throughout 2011 and 2012. In 2012, Microsoft unveiled the Surface, the first computer in the company's history to have its hardware made in-house. At the same time, Microsoft was buying the social network Yammer to compete with Facebook, launching the Outlook.com webmail service to compete with Gmail, and releasing Windows Server 2012 to compete with Amazon. In 2013 Microsoft agreed to buy Nokia's mobile unit for $7 billion, which was then followed in 2014 by the acquisition of Nokia Devices and Services, forming Microsoft Mobile Oy, and the acquisition of Skype. The Alliance for Affordable Internet was also launched in 2013, with

---

[23]In this sense, it should be emphasized the partnership Google keeps with Tag Heuer in the development of high quality smart watches and wearables in order to be able to compete with Apple's range of high quality goods.

Microsoft as part of a coalition of public and private organizations that also includes Facebook, Intel and Google, and aims to make Internet access more affordable so that access is broadened in the developing world.[24]

Yahoo has various other services besides the original search engine and Email (e.g., Yahoo News, Yahoo Mobile, Yahoo Messenger; Yahoo Music, Yahoo Finance, Yahoo! 360°, Flickr).[25] Yahoo has also signed partnership deals with different broadband providers such as AT&T, Verizon Communications, Rogers Communications and British Telecom, offering a range of free and premium Yahoo content and services to subscribers. In 2013 Yahoo purchased the blogging site Tumblr for US$1.1 billion, which led to a significant shift in the activities of the firm, targeting to get into the social network activities already in place in the other GAFAs. In 2014, the firm announced its partnership with Yelp Inc. and the acquisition of BrightRoll so as to compete with Google.

Amazon product lines include a terrific diversity of goods and services, from media to baby products, and jewelries to groceries. Being mostly recognized as a book retailer, it has to be said that the firm also counts with its own publishing unit. In 2011, Amazon announced its entry into the tablet computer segment in a move to get closer to the activities of the other GAFAs, who as discussed were already present in this segment. This launch was followed by the Amazon Appstore for Android devices. In 2012, Amazon announced it would be adding a gaming department (i.e., Amazon Game Studios) to get into the entertainment and gaming industry. In 2014 Amazon announced its Amazon Fire TV set-top box system, a device targeted to compete with such systems like Apple TV or Google's Chromecast device. On the one hand it allows for streaming videos from sites like Amazon's own streaming service as well as others such as Netflix or Hulu, while it also supports voice search for movies and games on the other. This should be interpreted as part of the interest of the firm in its Amazon Studios, a division focused on the development of TV shows, movies and comics. Next, the company entered the smartphone market with the release of the Fire Phone. Amazon has also carried out significant moves in the so-called Amazon Web Services (AWS), particularly as regards cloud computing and storage. In the past years, the company is also investing in the use of unmanned drones to deliver small packages and also 3D printing.

Facebook filed for an initial public offering in 2012, getting the largest valuation to date for a newly listed public company. At the same time, Facebook announced App Center, a store selling applications that operate via the site. Besides, it also acquired the firm Instagram and entered cloud storage. In its move toward the search engine business, in 2013 Facebook announced Facebook

---

[24]We should not overlook the fact that in August 2015, a consortium of major German automotive business (including Daimler, BMW and the luxury division of Audi and Volkswagen) has agreed to buy maps of the Finnish company Nokia for a value of €2500 million, in an attempt to expand the participation of auto manufacturers in digital online services. It is expected that these systems will have a key collision detection and other functions in driverless vehicles.

[25]See: http://www.diffen.com/difference/Google_vs_Yahoo. Accessed 10 August 2015.

Graph Search. Facebook also unveiled Facebook Home, a user-interface layer for Android devices, which were first made available in smartphones by HTC. In February 2014, Facebook announced the acquisition of the mobile messaging company Whatsapp for US$19 billion, which was followed by the acquisition of Pryte, a Finnish mobile data-plan firm that aims to make it easier for mobile phone users in underdeveloped parts of the world to use wireless Internet apps (see The Alliance for Affordable Internet above), the investment in the future of wearables through the acquisition of Fitness App, and the acquisition of LiveRail, an online video advertising company.

Samsung Electronics has emphasized innovation in its management strategy since the early 2000s. In the first quarter of 2012 the company became the highest selling mobile and smartphone company. These large earnings allowed the firm to (radically) get into different streams of activity (e.g. LCD and LEDs, semiconductors, Wi-Fi standards, Internet TV, connected housed and Internet of Things, wearables, virtual and augmented reality). In part, the fact that Samsung has become the world's biggest semiconductor chip supplier can be attributed to this financial success. In 2014, Samsung partnered with Amazon to introduce the Kindle for Samsung app. As most of the previous GAFAs, Samsung also counts with its own app store. While many other handset makers tended to focus on supporting one or two OS, Samsung kept supporting a wider range, like Symbian, Windows Phone, Linux-based LiMo, and Samsung's proprietary Bada.

Despite Twitter is one of the smallest of these Internet Giants, it has experienced very rapid growth. In 2012, Twitter acquired Vine, a video clip company that allowed users to create and share six-second looping video clips. In 2013, Twitter launched a music app called Twitter Music for the iPhone. In 2014, the firm announced the acquisition of Namo Media, a technology firm specializing in native advertising for mobile devices. This was followed by the acquisition of SnappyTV, a service that helps edit and share video from television broadcasts, and the acquisition of CardSpring, which enables retailers to offer online shoppers coupons that they can automatically sync to their credit cards in order to receive discounts when they shop in physical stores. More recently, and due to the increasing relevance gained by Meerkat, Twitter announced its acquisition of Periscope, an app which allows live streaming of video. Twitter is also increasingly used for making TV more interactive and social, not only for the audience but also for the TV companies themselves. In May 2013, it launched Twitter Amplify—an advertising product for media and consumer brands, and more recently it is also engaging into financing and banking. In an attempt to compete with Twitter's leadership in TV, Facebook introduced a number of features in 2013 to drive conversation around TV including hashtags, verified profiles, and embeddable posts.[26]

---

[26]This competition between Facebook and Twitter is increasingly being regarded as the "news war" (Holmes 2015).

So, which are the dominant vectors that the GAFAs are engaged in and which could reach mass consumer markets in the next years to come? Improving the efficiency of the terminals and devices implies accelerating the diffusion of the technologies and the devices that make up the Internet universe. However, it is hard to know with certainty what the dynamics in the efficiency improvements of these devices might be, and the timing in which these can be achieved. Computers and mobile devices (e.g., smartphones, tablets) are progressively introducing technological improvements in the new generations through advances in batteries (i.e., improving battery life and speed of connection), efficiency gains and price. For example, screens, larger and with much better definition, are demanding more energy and requiring superior performance, so it is necessary to adjust battery consumption. In this regard, graphene presents ideal properties, with respect to silicon, ceramics and plastics, to manufacture components that can later be incorporated into the devices of the future. In fact, Asian firms Moxi and Galapad have already announced their idea to launch in short 30,000 devices in which graphene is used in batteries, screens and power systems.

Competition between the various agents that form the Internet constellation, also results in the development of time-varying dynamics. As we have earlier discussed, it is unpredictable to know in advance and with certainty the evolution that the different groups and actors may have, due to the clashes and conflicts among them.[27]

Wearables and smart watches have significantly grown in these past years, which have not only become an important market for the GAFAs, but are also transforming sectors such as health or sports. The so-called "phablets" (i.e., a hybrid between smartphones and tablets) are also growing rapidly. Although this can be regarded as an incremental move, according to the forecasts made by Business Insider (Danova 2014), sales of phablets in 2019 will triple that of tablets. The increase in the demand of these phablets is due to the expected increase in the generation and further use of services and content, together with the and growth in advertising investment in mobile devices.

The level reached by the technology, the ability to connect appliances, and particularly the progress made in human-machine interaction makes it possible to contemplate the potential of the Internet of Things. The development, deployment, and distribution of Internet of Things requires a very important network of telecom support and infrastructure. This explains the introduction of telecom operators in areas such as 4G and 5G (see Chapter "4G Technology: The Role of Telecom Carriers" by Araujo and Urizar), what in turn leads to joint ventures and mergers or acquisitions. As with mobile payment, new alliances and partnerships are being established among different players like General Motors with AT&T, or Sprint Nextel with Chrysler. But as it is the case with any connection, the Internet of Things is still vulnerable and subject to risks of cyberattacks, so here too, it is not easy to accurately assess the speed of this phenomenon.

---

[27]For example, competition between providers of processors and chips (key components of smartphones and wearables) is becoming increasingly noticeable.

The Internet of Things is conducive to several global trends; for example in relation to the consumption of food, health and fitness, besides the already mentioned wearables. While the arrival of driverless cars is still pending, we are witnessing a race to connect cars and other physical objects to the Internet and integrate apps services in these. Manufacturers, operators and technology companies have started the conquest of this new market, but, again, there are still major obstacles to overcome.

Mobile payments are still modest, but it is estimated that we are close to reaching a context in which cash money ceases to have relevance, and economic transactions can be made through mobile terminals safely.[28] Technically, there are different solutions available; but this has not yet being enough for a sharp growth in mobile payments due to the different and sometimes conflicting interests of the various stakeholders involved: telecom operators, mobile device manufacturers, banks, etc. In this regard, the entry of Google, Amazon, Apple, Twitter, or Facebook, in mobile payment is becoming an important competition for traditional banks.

Partly due to the increase in online shopping, there is a trend in many cities to a gradual disappearance of the traditional shops and local commerce. Globalization trends and the potential of the information made available through the Internet are posing a major challenge to these small and much localized actors. In this sense, large groups are increasingly offering these small businesses the opportunity to be visible through marketing techniques on the net (see the Ads provided by many of the GAFAs), so their scope of activity can adapt to the new globalization requirements. The universe of the Internet is in present times characterized by the existence of massive or big data. As we have discussed earlier, Amazon was the first among the GAFAs to get into the big data through the Amazon Web Services. The analysis of this massive data requires the involvement of specialized service companies, what explains the increasing moves of the GAFAs toward cloud computing activities and services (see also Chapter "The Digital Ecosystem: An "Inherit" Disruption for Developers?" by Vega et al.). The advice to local companies and ships is thus regarded crucial so these can benefit from the opportunities that arise from the big data. As a result, large business groups must be able to provide customers with services that adhere to their main specialty. For example, large groups such as the Spanish bank BBVA are redefining their strategies toward the provision of software services and the transmission of data (Gallego 2015). Competition between the GAFAs groups is occurring particularly between Microsoft, Facebook, Amazon, and Google, with unpredictable results in the medium term.

Increasingly, companies and countries are in an international context of cyber-attacks, in which every organization can be subject to being attacked. Table 6 offers an illustration, though not exhaustive, of the cyber dangers affecting the potential GAFAs' dynamics in the short term.

---

[28]Mobile payment requires the deployment of Near Field Communication technology, which is still widespread in very few countries.

Table 6 Cyber challenges and dangers faced by the GAFAs and other SMEs

Malware	Short for "malicious software" is any type of program designed or used for unauthorized access to a computer system
	It can be used to access data, control a targeted system, or to do both. Malware used to access data ranges from simple programs that track keystrokes and copy screenshots to sophisticated programs that can search through a users files and browser history to steal passwords and bank data
	While malware has historically targeted only computers, "mobile malware" that targets tablets and smartphones is an increasing threat
	Terms often used in security news stories like viruses, worms, Trojans and spyware describe specific types of malware
Phishing	Phishing attacks have become more sophisticated in recent years as the online footprints of individuals have grown. Social networks have given phishers access to a treasure trove of personal information they can use to customize their attacks and increase their likelihood of success
Phishing scams	Phishing Scams may use email, text messages, Facebook or Twitter to distribute links to malicious webpages designed to trick you into providing information like passwords or account numbers
Advanced persistent threats	Are systematic, long-term attacks against technology systems. They seek to create situations for very complex malware programs to be introduced or permitted access to critical systems or information. They involve the accumulation of several strategies, including: phishing, social engineering, waterholes, or exploratory hacking to mention a few
Unsecured internet connections	Businesses do not have direct control over these wireless access points like they do in the workplace, and these unsecured connections risk exposing company data when security measures are not taken to protect the transmission of data
Cloud computing	In spite of the risks associated to cloud computing, the resources devoted to this area on cyber security are still quite low
Passwords and encryptions	Hackers can use special software to "guess" passwords or they can trick unsuspecting employees into turning over their login credentials by directing them to seemingly legitimate login pages
Application-based threats	So-called "malicious apps" may look fine on the surface, but they are specifically designed to commit fraud or cause disruption to devices. They may come in the form of malware, but also include privacy threats and vulnerable apps
Phishing scams	May use email, text messages, Facebook or Twitter to distribute links to malicious webpages designed to trick you into providing information like passwords or account numbers

Source Own elaboration based on Harris (2014) and Hobby et al. (2014)

Table 7 Possible measures to face the cyber risks

The "Leaders"	Understand and be conscious about the company's current exposure to cyber-threats and its effectiveness in managing the risk
Employees and managers	Should follow good practices of cybersecurity, and define crisis plans against possible attacks
Predefining recovery point objectives	They represent the maximum acceptable data loss, the maximum tolerable time data, services, and operations can be unavailable, as the result of an incident
	Since an attack transforms the environment, it would be possible to assess the resilience to these drastic events
Firms	For more details on the 21 guidelines organizations may follow to protect their core business, see Hobby et al. (2014: 12)
Prevention systems	More comprehensive encryption systems should be implemented, together with antivirus programs and appropriate firewalls, by specialized suppliers
Cloud computing service providers, contract compliance providers, and mobile application providers	Firms should negotiate with these the potential problems and security risks to be faces, so as to choose the best supplier
Services provided by telecom operators	Besides providing access to high speed capacity (4G, 5G), they should also guarantee security in their networks

Source Own elaboration based on Harris (2014) and Hobby et al. (2014)

Epigenetic dynamics are directly influenced by the dangers outlined in Table 6. This implies that the evolution of these dynamics might be faster (i.e., more disruptive) or slower (i.e., more incremental and progressive). Thus, a high degree of uncertainty and unpredictability is introduced into the system to predict how and to what extent (i.e., speed) nuclear technologies to the further development of the Internet ecosystem can spread, what affects both the GAFAs and their users. Next, we present some of the practical steps or roads organizations and individuals may follow to minimize cyber vulnerabilities, and defend from possible attacks and the challenges associated to them as discussed earlier. The result of the possible ways to meet the challenges below by companies is not predictable. In certain contexts, the difficulty of overcoming these challenges will be greater and in others instead easier to address. Therefore, anticipating and predicting the speed of certain evolutionary dynamics is not possible, unless the characteristics of the environment in which the organizations under analysis are embedded are known (i.e., stage one of the EED approach). However, what it can be concluded is that without the measures outlined in Table 7, any change in the environment (e.g., changes associated to the growth in demand for Internet users) would be more constrained, and even canceled, and therefore any potential path resulting from the new epigenetic dynamics may vanish.

5 Patenting Dynamics

As we have earlier discussed, patents are one of the strongest environmental properties in of the Internet ecosystem. The field of patents shows just how fierce the competition is. The main reason is the high knowledge content in these groups' business activities. When the GAFAs enter new fields, they embark on out-of-control patent purchasing (a scenario known as the "patent war"). Examples of the increased dynamics of acquisition and penetration in new fields can be the seventeen thousand Motorola patents bought by Google, the CPTN Holding, formed by Microsoft, Apple and other companies, which acquired six thousand patents auctioned by the bankrupt company Nortel, or Facebook's patent acquisitions from IBM and AOL (Gómez-Uranga et al. 2014). As a result, the groups get involved in an enormous amount of cross litigation, to support their business strategies aimed at "doing new business" or protecting themselves. However, we have not included here the analysis of the patents acquired as a result of the joint ventures or acquisitions of other firms, as the analysis of the M&As is carried out in the next section.

In this section, we will illustrate whether the previously observed epigenetic dynamics can be explained by the patenting behavior of these business groups. The goal is to analyze the evolution and dynamics observed in the GAFAs, determining how these firms expand and diversify their activities. With it, we aim to analyze the behavior of the groups in their patent portfolios to illustrate the EED approach.

Figure 19 illustrates the total number of USPTO patents granted to each of the GAFAs under study between 1984 and 2014. Naturally, the exponential differences among these can be partially explained by the age of the firm. While Twitter, the youngest firm among all, was born in 2006, Microsoft and Samsung Electronics go back to 1975 and 1988, respectively. However, the voluminosity of the patents also provides explanations for the epigenetic dynamics discussed above.

In Sect. 3 (see Fig. 9), we provided evidence of the number of patents granted per year at the USPTO to the GAFAs between 2007 and 2014. Here, we will go back to this figure so as to analyze more in-depth the voluminosity of the patenting activities in the GAFAs. As the reader may observe, we have divided Fig. 20 into two blocks. Part a focuses on the evolution shown by the five business groups with the largest amount of patents granted in the past 7 years (i.e., Google, Apple, Microsoft, Nokia and Samsung Electronics), while part b focuses on the five groups with the smallest number of granted patents (i.e., Facebook, Yahoo, Amazon, Twitter, and eBay). The first conclusion that can be obtained from the figures below is that Samsung Electronics is the company that a priori, has the strongest technological capabilities to lead potential epigenetic dynamics in the next years. The 5794 USPTO patents granted in 2014 provided the firm with a large technological advantage in the development of LCDs and LEDs, semiconductors, Wi-Fi standards, Internet TV, connected housed and Internet of Things, wearables, and virtual and augmented reality products. Samsung is also the firm that counts with the largest accumulative knowledge in the protection of

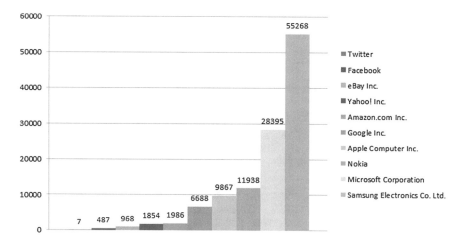

Fig. 19 Number of USPTO patents granted between 1984 and 2014. *Source* Own elaboration based on USPTO

its know-how as the 55,269 USPTO patents granted between 1984 and 2014 evidence. Besides, it should be stressed the evolution shown by the company in this very matter in the past 7 years. While in 2007 the firm was granted 3332 patents at the USPTO, this amount was almost doubles in 2014, an evolution that only Apple and Google seem to be able to follow. Microsoft and Nokia have on the other hand remained somewhat constant over these past years, with values close to 3000 and 900 patents, respectively.

When we move the analysis to the remaining five GAFAs, the first thing to note is the difference in the values observed in the vertical axis. Where in Fig. 20a, we were close to 6000 patents per year, in here it is Amazon the one that shows the largest number of patents granted per year, with 751 patents in 2014. The second aspect that should be noted, though in different scales, is the parallel path followed by Amazon and Facebook. Facebook got its first patents granted ($n = 7$) in 2010. The numbers were kept constant in 2011 with eight patents, but since then the evolution has been explosive with 50, 131, and 291 patents in the years 2012, 2013, and 2014, respectively. As it was the case with Microsoft and Nokia above, the path followed by Yahoo since 2010 can also be regarded as stable, with around 300 patents per year. Finally, eBay has also shown progressive growth rates since 2010 in particular, while the very low values achieved by Twitter to date, with three patents in 2013 and four in 2014 do not allow reaching any solid conclusions.

In order to analyze to what extent the dynamics followed by the GAFAs are more or less disruptive, and therefore more toward or away from the original DNA of these groups, we have next conducted a partial analysis of the diversification paths followed by the GAFAs in their patenting strategies. In order to achieve this goal, we have gathered all the (CPC—cooperative patent classification) technology classes included in the patents granted to the GAFAs at the

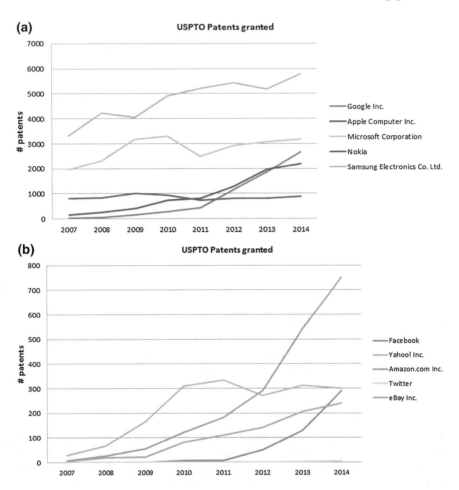

Fig. 20 **a** Number of yearly USPTO patents granted to some of the large business groups operating on the Internet (2007 2014). *Source* Own elaboration based on USPTO. **b** Number of yearly USPTO patents granted to some of the large business groups operating on the Internet (2007–2014). *Source* Own elaboration based on USPTO

USPTO. In each case, this analysis has been carried out since the year in which the first patent was granted to each of the 10 GAFAs included in the chapter. Once all technology classes covered by the patents granted each year are known, then we have assessed which is their share in relation to the total number of patents for that year.[29] In other words, we measure which is the number of times a certain technology class is repeated as compared to the total number of patents

---

[29]The technological domain covered by each technology class can be observed at the World International Patent Office (WIPO). See: http://web2.wipo.int/classifications/ipc/ipcpub/#ref resh=page&notion=scheme&version=20150101&symbol=G06F0017300000. Accessed 19 December 2015.

Table 8 Methodology followed to assess the technological diversification of the GAFAs. *Source* Own elaboration

Year 1		
Tech class (CPC)	# patents	%
	2	3
G06	3	100.00
Y10	2	66.67

in that particular year. Let us assume a company has been granted 3 patents in a year, and these three patents include two technology classes. We observe the number of times each technology class is being included in the 3 patents and from there we are able to assess the relative weight of each technology class. This operation is then repeated for all the 10 GAFAs and for all years in which they have been granted patents at the USPTO. However, in order to provide a preliminary analysis of the results obtained so far, we will only represent below the diversification paths followed by three of these firms, namely, Twitter, eBay, and Amazon.[30] The vertical axis represents the different technology classes in which each of the groups has obtained patents in their evolutionary paths. In turn, the size of the bubble represent the share of each technology class in relation to the total number of patents (i.e., the % as outlined in Table 8), so that the larger the share, the bigger the size of the bubble.

As indicated above, Twitter got their first three patents granted in 2013, to reach a total of seven patents to date. These seven patents only include two CPC technology classes, G06 and H04. As Fig. 21 shows, the relative weight of the technology class G06 has remained constant over time, while the weight associated to H04 has increased from 33.33 % in 2013 to a 50 % in 2014.

In the case of eBay, the company got its first patent granted in year 2000, so in order to gain some conclusions about its technological diversification, the time window is much larger in this case. As Fig. 22 shows, eBay too started patenting two technology classes (i.e., G06 and Y10), a situation that remained constant until the year 2004 when a third technology class was included (i.e., H04). These three technology classes have ever since remained in eBay's patent portfolio. It is noteworthy signaling year 2009 as the year in which more technology classes started being included in the know-how of the firm. Since then, the diversification has continued including more classes to the technological capabilities of the firm. Even if this diversification is clearly observed, however, the relative weight of these classes in the whole portfolio of patents has remained fairly constant, with three technology classes dominating above the others (i.e., G06, H04, and Y10).

---

[30]Another reason for not including here the evolution of the other firms is visibility. As the reader may expect, the larger the voluminosity of patents, there bigger also the number of technology classes covered. For example, Samsung Electronics had 77 different technology classes only in year 2014, Nokia and Microsoft 31, Apple 54 and Google 39. If we add these numbers to the values for the technology classes in the rest of the years, we end up in situations where more than 200 classes need to be visualized, which we deem not sensible to provide in one single figure.

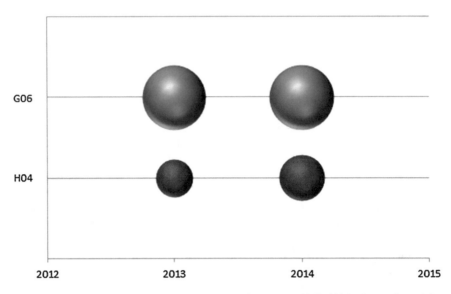

Fig. 21 Technological diversification path followed by Twitter (2013–2014). *Source* Own elaboration based on USPTO

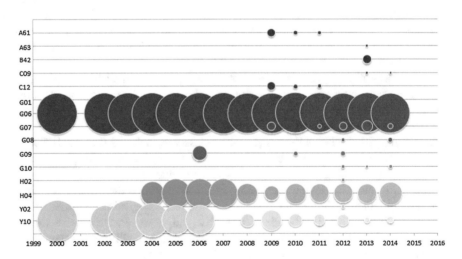

Fig. 22 Technological diversification path followed by eBay (2000–2014). *Source* Own elaboration based on USPTO

The previous dynamics are also replicated in the case of Amazon but to a much larger extent. In other words, the dynamics are more disruptive. As Fig. 23 shows, Amazon's portfolio of patents incorporates many additional technology classes than in the previous two cases. Besides, the relative weights of these additional classes are also much larger than in the previous cases. We can for example refer

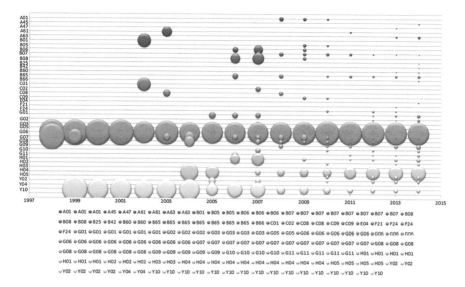

Fig. 23 Technological diversification path followed by Amazon (1998–2014). *Source* Own elaboration based on USPTO

to classes such as A61, B01, B06, B08, C01, G01, G07, and H01 which are also representative of the dynamics followed by Amazon in recent years.

So, what can we say about the research questions posed at the beginning of this chapter? In relation to what technological areas are the GAFAs moving into overtime, the previous analysis shows that the high velocity of the Internet ecosystem is making the analyzed firms move into other technological classes which were not at the core of their organizational routines. Some of these are: A63, B41, B60, B82, G01, G08, G09, G11, H01, H03, and H05. Are they related to these groups' DNAs? Addressing this research question is a bit more complex, since the technological capabilities of each firm varies, and therefore it is not possible to provide a unique answer. In order to be able to comprehensively assess this question, it would be necessary to carry out an analysis of the technological distances among technology classes, similar to these introduced by the scholars in the field of economic geography as regards related variety studies (see Sect. 1). However, what it can be confirmed is that fact that these diversification dynamics are similar across the GAFAs. Even if the intensity of the multiple technology classes across firms varies, when we observe the diversification paths followed by all GAFAs is parallel. Thus, as we have discussed in the book, the environment is forcing all the firms to go through similar paths. This causes contexts in which all go against all, as previously discussed in the chapter. The main rationale for this patenting war is that as Malik (2015) evidences, and as Abba would sing, in the context of the Internet ecosystem, the winner almost always takes it all.

This analysis has also driven us to suggest the following research avenues and pursue the following research goals. Given the large amounts of patents granted to the GAFAS, we believe it would be very enlightening to apply the latest methodological approaches developed in the scientometric and technology mining communities (van Eck et al. 2013; Leydesdorff et al. 2014; Kay et al. 2014) to gain new insights on their dynamics and their diversification strategies. This would pursue the following goals:

– The analysis of the patents granted to the GAFAs will allow us to bundle all the patent information into clusters of activity, so that the main dominant vectors in which the GAFAs diversity their portfolio of activities are identified. With it we will be able to determine which are the key milestones (back in time) that set the ground for the diversification that we are witnessing at present times. The previous analysis will also allow us to determine whether the GAFAs are following related variety or unrelated variety strategies in their diversification processes. This analysis may provide crucial information not only for other established firms in a large variety of industry segments who want to supply these large players into their diversification strategies (e.g., security, health, financing), but also for new ventures (i.e., entrepreneurs) who are identifying technological and market opportunities.
– The analysis of the citations of the granted patents (backward and forward citation) can help us understand which firms cite the patents of the GAFAs and which patents are cited by the patents granted to them. Similarly, we will also be able to examine the extent to which the GAFAs cite each other's patents or not.
– The analysis of the citations of the patents granted to the GAFAs can also provide relevant information for the identification of which the standard essential patents are (Gómez-Uranga et al. 2014). By studying the forward and backward citations of the patents granted to the GAFAs we would be able to observe whether there are certain patents that play a central role in opening new technological domains. Besides, once these "essential" patents are identified, then it will be possible also to track who their inventors are as well as their affiliation and location (at the moment of patenting).
– As regards the analysis of the key inventors, we could also identify whether these only cooperate with one company or whether they patent for many companies at the same time. Inventors can also provide relevant information as to the extent to which they are present in many different communities at the same time, so the radical innovations are produced as a result of multidisciplinarity. In this regard, some of the questions that could be addressed from the perspective of the inventors are: Which are the communities these lead inventors are present in at each moment in time? And how many communities are they present in at the same time? Why do researchers/inventors move between often unrelated communities? How distant are these communities?

In order to achieve the previous research purposes, it would be possible to rely on two well established methodologies in the innovation studies community. On the one hand, social network analysis methods are applied (Granovetter 1973; Håkansson 1987; Ahuja and Katila 2001). Given the large amounts of patents to be analyzed, by studying the forward and backward citations it is also be possible to answer identify those patents that act as standards in the mobile and software industries. By identifying the patents that act as standard essential patents, it would also possible to understand the reasons for the GAFAs to engage in patent lawsuits one another.

6 Merger and Acquisition Dynamics

In this section, we will focus on the number of M&As completed by each of the GAFAs, and the amount of investment required in order to explain the identified epigenetic dynamics. As Daidj (2011) discusses, the motivations for carrying out M&A include achieving growth by opening up to market opportunities in domestic and foreign markets; having a better access to capital, intangible assets of other firms such as managerial skills and knowledge of markets and customers, etc. In the context of high velocity environments as the one we are interested in, firms pursue M&A to renew their technical capabilities and products. However, M&As are regarded as one of the most effective ways to spur innovation and change the markets in which the firm is either competing or aims to compete.

The GAFAs have for long signed agreements with different partners belonging to the ICT sector but also to the automotive, banking industry, etc., particularly in these past years where their number and intensity has bubbled. To a certain extent, we could make a metaphor here with the GAFAs and the pirates. Pirates were looking for the best of the best anytime and anywhere, and similarly, the GAFAs are very much aware of their antennas in order to capture the innovations available worldwide and bring them to their own "castles." Figure 24 shows the total number of M&As completed by the GAFAs between 1987 and 2015. As it can be

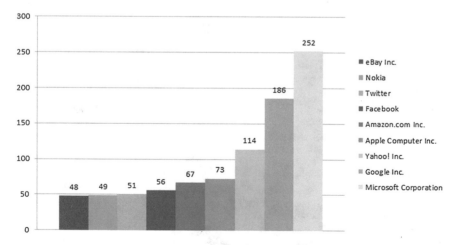

Fig. 24 Number of M&As completed (1987–2015) (We could not find any evidence of the M&As completed by Samsung.). *Source* Own elaboration

observed, according to the total number of M&As, Microsoft is the one that counts with the largest portfolio of firms with 252. However, the case by Google deserves special attention, because despite it is a much younger firm than Microsoft outperforms all the other organizations in this dimension. In this particular case, M&As are without doubt one of the main factors explaining the epigenetic dynamics and hence, the high adaptability shown by Google.

Next, we plot the total accumulated value (in million US$) of the M&As completed by the previous firms over time. In this case too, the company that has invested the most in the acquisition of firms is Microsoft with US$52,124 million, followed in the distance by Google with US$27,789 million. If we compare the two figures, the reader may note that Facebook and eBay have increased its relative position with regard to the value invested in the acquisition of other firms. On the contrary, the case by Apple is worth stressing as it is one of the firms that has invested the least in acquiring new firms (US$6211 million). It has to be noted that the values represented below, only include those acquisitions whose value has been reported and does not remain undisclosed. As a result, due to the large number of acquisitions for which we could not find information, Fig. 25 should be interpreted with caution.

Figure 26a, b we represent the evolution shown by the GAFAs with regard to the number of M&As. As we did in the case of the analysis of the patents in the previous section, here too, we divide Fig. 26 into two blocks. Part a focuses on the evolution shown by the five business groups with the largest number of M&As (i.e., Google, Apple, Microsoft, Nokia, and Yahoo). In the case of Google, the number of acquisitions, taking into account the evolution of the firm, shows peaks in 2007 with 15 acquisitions, 2010 with 27 acquisitions, and 2014 with 32 acquisitions, respectively. Microsoft also shows various peaks in years 1999, 2006 and 2008, and 2015 with 32, 18 and 19 acquisitions, respectively. As discussed earlier, Apple is not characterized by the completing a significant number of M&As. The

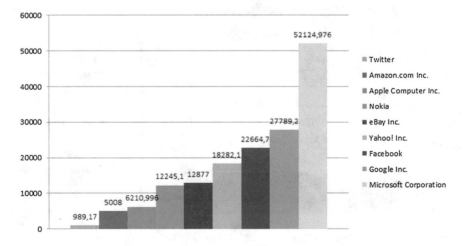

Fig. 25 Value of the completed M&As (1987–2015) (Million US$). *Source* Own elaboration

number of M&As has been kept rather stable between 3 and 4 yearly, until year 2013 when the company acquired up to 14 firms. Finally, Yahoo also shows two peaks in the number of MA&s, the first in 2005 with 10 M&As and the second in 2013, just one year after Marissa Mayer was appointed as the CEO of the firm, with 27 firms acquired. In turn, part b of the figure focuses on the 4 groups with the smallest number of acquisitions (i.e., Facebook, Amazon, Twitter and eBay). As can be observed, these four firms start conducting M&As later in time as compared to the other five cases. Amazon and eBay conducted their first M&As in 1998, while Facebook completed its first acquisition in the year 2005 and Twitter in the year 2008. Among these, the case of Amazon is quite illustrative of a first mover. The company, born in 1994, started acquiring other firms just 4 years after

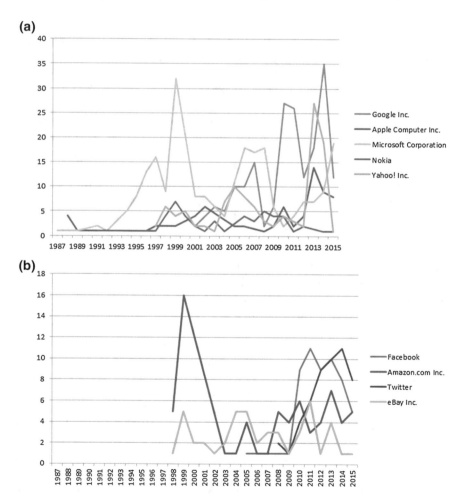

Fig. 26 **a** Number of M&As completed (1987–2015) by some of the large business groups operating on the Internet. *Source* Own elaboration. **b** Number of M&As completed (1987–2015) by some of the large business groups operating on the Internet. *Source* Own elaboration

its existence, and in fact, the largest volume of acquisitions was observed in 1999 ($n = 16$). In the case of eBay, the number of acquisitions shows ups and downs with values in between 3 and 6 firms acquired yearly. Finally, both Facebook and Twitter show a parallel path with an exponential increase in the number of firms acquired.

In Fig. 27, we replicate the previous analysis but this time with the value invested in the M&As instead. Here too, we divide the figure into two blocks. Part a focuses on the evolution shown by the five business groups with the largest investments devoted to the acquisition of new firms (i.e., Google, Microsoft, Nokia, Facebook, and Yahoo), while part b focuses on the 4 groups with the smallest number of granted patents (i.e., Apple, Amazon, Twitter, and eBay).

As to the former, the most illustrative cases are the peaks observed for Yahoo in 1999, Nokia in 2008, Google in 2011 and Facebook in 2011. In the case of Yahoo, the company invested in 1999 a total amount of US$9510 million in the acquisition of four firms, the most significant of which were Broadcast.com for a value of US$5700 million, and Geocities for US$3600 million. In the case of Nokia, the Finnish company invested US$8517 million in year 2008 in the acquisition of five firms, from which the acquisition of the American company Navteq is to be stressed with a value of US$81,000 million. In 2011, Google devoted US$13,265 million to acquire 26 firms. Among these firms, the most significant ones were Motorola for a value of US$12,500 million, a case that has already been discussed in Gómez-Uranga et al. (2014). In turn, Facebook invested in year 2014 US$21,500 million in the acquisition of 35 firms, the most relevant of which were well-known cases of Whatssap for US$19,000 million and Oculus VR for US$2000 million. Finally, and for Microsoft five peaks can be observed in years 1999 (32 firms for US$7116 million), 2005 (11 firms and US$3175 million), 2007 (17 firms and US$6793 million), 2011 (US$8600 million in four firms) and 2013 (US$7200 million in seven firms). Among these, the most significant acquisitions were the ones of AQuantive, a digital marketing firm in 2007 for a value of US$6333 million, the acquisition of Skype in 2011 for US$8500 million, and the previously discussed acquisition of Nokia mobile phones unit in 2013 for US$7200 million.

In the case of Apple, the most substantial acquisitions were completed in year 2014, when a total amount of US$3030 million were invested in nine firms, the most significant of which was the music streaming company Beats Electronics (US$3000 million). In the case of eBay, there are four years which deserve attention, 2002 (US$1500 million in one firm, Paypal), 2005 (US$3220 million in five firms), 2008 (US$1759 million in three firms) and 2011 (US$2825 million in six firms). In Amazon, there are three peaks that deserve some attention in years 2009 (US$1200 million in four firms), 2012 (US$1087 million in four firms) and 2014 (US$970 million in four firms). Among these, the most significant one was the acquisition of the online shoe and apparel retailer Zappos in 2009 for US$1200 million.

Here too, there are several further research avenues to be conducted in order to better understand the role M&As play in explaining the epigenetic dynamics of

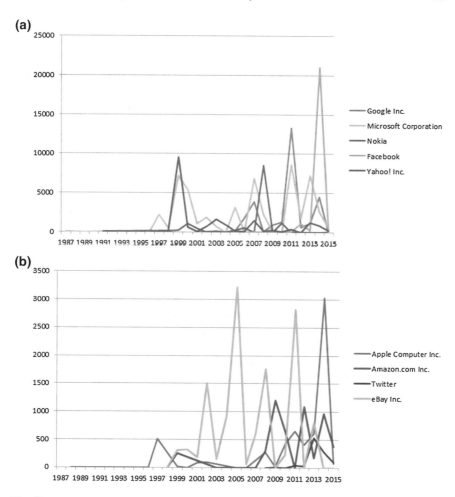

Fig. 27 a Value of the completed M&As (1987–2015) by some of the large business groups operating on the Internet (Million US$). *Source* Own elaboration. **b** Value of the completed M&As (1987–2015) by some of the large business groups operating on the Internet (Million US$). *Source* Own elaboration

the GAFAs. Which domains of activity do the acquired start-up firms belong to? Does the diversification strategy, through the acquisition of new firms, respond to a related or an unrelated variety? What is their geographical location? Are they located close to the GAFAs (i.e., hot spots) or are they dispersed? How old are these firms? There are still many open questions to be addressed and that we have not yet been able to respond in a comprehensive manner.

The analysis of the age, size of the acquired firm, its geographical location, and the sector it belongs to will allow us to conclude whether the GAFAs are following related or unrelated variety strategies. Besides, since many of the entrepreneurial firms that are acquired by the GAFAs also count with an important patent portfolio, the questions formulated above for the case of the GAFAs could also be pursued, but in relation to the acquired entrepreneurial firms created by developers.

One of the characteristics of modern developers is the speed at which they can create new applications, programs, etc. In this regard, the cloud, the use of automation tools, collaborative methodologies, ready-made components, the availability of open software and code and the large number of developers facilitate their diffusion into the Internet ecosystem. Developers get adapted instantaneously to:

- The relationship with other companies which either absorb their applications (e.g., the GAFAs) or use them in their respective markets and sectors (e.g., machine tools).
- The requirements and limitations of the hardware through the development of new programs (software) or changing and improving existing ones.
- The areas (i.e., dominant vectors) where a higher demand for applications exist, such as the sustainability of the environment, health, human mobility, massive online shopping, videogames, etc.

However, in spite of the relevance of developers, there is certainly no evidence to date on their evolution, the dominant vectors that guide their actions, the reasons for the developers to orient toward certain industries ignoring others, etc.[31]

Building upon the data gathered in relation to patents and M&As, it would be possible to build a model that helps determine which are the key structural characteristics that explain: (i) the revenues achieved by the GAFAs, and (ii) their market capitalization. In this regard, a longitudinal model could be built, which would use panel data from a large variety of indicators characterizing the GAFAs, including: number of employees, gross profit, R&D investments, number of USPTO granted yearly, etc. As a result, we could identify which are the key variables that help statistically explain the limitless performance of the GAFAs.

7 Discussing Some of the Consequences as a Result of Epigenetic Factors

As Gómez-Uranga et al. have discussed in Chapter "Introducing an Epigenetic Approach for the Study of Internet Industry Groups" in this book, the third stage deals with the abnormalities, malfunctions or obstacles to innovation, and/or blockage to developing competition that may arise as a result of the influence of epigenetic factors (Lehman and Haslam 2013). Some of these implications include (see Gómez-Uranga et al. 2014): existence of a gap between R&D investments and patenting results; patenting rationale distorted; excessive transaction (and

---

[31]See Chapter "The Digital Ecosystem: An "Inherit" Disruption for Developers?" by Vega et al. for a methodological proposal on how to address the study of the dynamics of developers.

litigation) costs; high entry barriers to SME patenting; problems in the definition and development of standards; overload in patent offices and regulating agencies. Other potential consequences also include: economic consequences, institutional consequences, regulatory consequences, social consequences, moral and ethical consequences, with their subsequent implications in terms of policy and social dynamics (Dutton 2013). An example, currently under discussion in Europe, is the increasing tax engineering practices of the GAFAS (Corkery et al. 2015).

Patents are one of the strongest environmental properties in the field where the business groups examined in this book operate. In theory, patents ensure progress and technology advances. In practice, they have become a battlefield for cross-claims which questions one of the key objectives of patents systems. Patents are now being used to hinder competitors' growth. The meaning of patents has changed (Bessen and Meurer 2008): they used to be the result of innovation and companies could pay for the use of license rights, but now they seek exclusive rights so as to include them in their ecosystems and thus hinder rivals companies' growth (i.e., blocking the potential innovation capacity of competitors rather than creating the necessary incentives to innovate).

Companies sometimes seriously alter competition through their lawsuits, asking the court to stop the sale of their rivals' products (Fischer and Henkel 2012; Knable Gotts and Sher 2012). These business groups devote large amounts of their resources to patents, above all, human resources. Swarms of engineers and attorneys work constantly on companies' legal claims and law suits. Maintaining a patent portfolio and licenses on the basis of lawsuits involves enormous costs (Gómez-Uranga et al. 2014). Lawsuits and legal disputes not only involve high costs but also create disincentives for innovators. Bessen and Meurer (2008), state that the intellectual property rights system has failed as a means of protection and information for companies in the U.S. Lawsuits for infringement of intellectual property may even be affecting different business groups' share prices. Although the situation of patents and incentives for innovation varies according to the industry, software patents are very abstract and poorly defined. This makes it much more complicated to achieve reasonably efficient market contracts (Bessen and Meurer 2008). Therefore, we could say that market failure is due to poorly defined property rights. All of these issues lead us to ask if patents systems can no longer fulfill their primary objectives.

Business groups have a distorted view of the competition's conditions as a result of the turbulence in these environments. Companies may block or have their innovation paths blocked. Their competitive conditions may vary radically due to changes in the environment. The environment is at the mercy of whatever company is the biggest at a given time (Gómez-Uranga et al. 2014).

As we have seen, patent applications for operating systems and programs used in all types of computation, internet and telephone devices and gadgets have increased sharply in the past years. This has led to a considerable increase in the amount as well as the voluminosity of patents. This growth in the voluminosity (size and scope of patents applications) may be a necessary condition for companies to adapt their business models to changes in technology, legal systems and market conditions (van Zeebroeck et al. 2009).

Furthermore, it is important to take into account the impact caused by inefficiencies in the patent system in regard to the high price of the end product/service as well as higher "transaction costs" resulting from patenting expenses and related lawsuits (Gómez-Uranga et al. 2014). This inefficiency also means that products/services take longer to reach the market because of the time involved in patenting and the lawsuits which may result. The inability to control planned obsolescence is another aspect related to patents systems. This is due to overprotection of previously patented products and the strong deterrent posed by a lawsuit on behalf of the owners of the intellectual property concerned. Finally, we have to take into account the regulations and laws protecting possible violations of free competition resulting from predatory business practices when companies, etc., use their power to gain control of large amounts of intellectual production (i.e., patents).

Understanding the consequences that result from the discussed epigenetic dynamics of the GAFAs requires a dialectical vision in which there are sets of consequences that go in different directions but exist simultaneously. One of the positive consequences of the evolution of the Internet phenomenon (based on the evolutionary dynamics of the large groups) would be the increase of individuals to access new information, what can potentially lead to new paths that make these individuals better off. These could be called as "user-centric epigenetic dynamics." Other positive consequences could also include: (i) the increase of relationships among individuals; (ii) these connections and relationships are reinforced regardless their training and education levels; (iii) cultural and language barriers that impeded relationships are broken down; (iv) the possibilities of promoting extensively and intensively leisure activities are increased (e.g., music); (v) individuals can get rid of corsets and restrictions in their respective environments, and gain access to different universes elsewhere.

Meanwhile, one of the main negative consequences of the evolution of the Internet phenomenon (based on evolutionary dynamics of the large groups) refers to the increasing disintegration of privacy (Dutton 2013). In this regard, an example of direct user privacy attack is illustrated by Max Schrems, whose lawsuit against Facebook has recently been resolved by European courts. However, the most important and significant risks, both due to their size and their subtlety, are generated by the use Internet groups make of users and their information (i.e., personal data) for economic and commercial purposes. Other potential negative consequences include: (a) changes in human relationships, what results in an impoverishment of these, despite their exponential increase; (b) relationships in the Internet world, even when they have a more horizontal character, pose no greater democratization of these; On the contrary, they often imply maintaining and reinforcing the status quo of the establishment, which are mainly represented by the interests of the large Internet groups; (c) a distortion of world economic resources, whose fate is not devoted to the improvement of the living standards of the general population; (d) a greater capacity to generate high risk situations, due to the possible loss of control over potential cyberattacks that could have devastating effects on economies, populations, etc.

European Commissioner Oettinger (Digital Economy Society) has as main responsibilities, among others, to take steps toward a connected digital single Europe and to achieve a comprehensive protection of data and copyright. The

Commissioner noted that Europe must force business groups to comply with the rules issued in Europe. Clearly the past stances of European Commissioners during the past years open a particular conflict, not without tensions between the U.S. and European governments (Leyden and Dolmans 2014; Wiethaus 2015).

In any case, in recent years, we are witnessing the existence of a gap between the evolution of technologies on the one hand, and the activities and practices of Internet groups and the adequacy of regulations and laws to these changes on the other. The latter always move with a delay. One of the latest examples in this regard is the regulation on network neutrality issued by the Federal Communications Commission in February 2015 in order to guarantee the promotion of the quality of Internet service (Tardiff 2015). Against radical solutions such as the rejection or the submissive acceptance of judicial decisions, or the very high fines, other intermediate areas, based on negotiation and consensus, and in finding policy approaches that may better suit to a situation of conflicting interests are imposed (Nazzini 2015).

The dominant discourse on technology provides the latter with an almost total autonomy on issues such as culture and society. In this "standard" understanding of technology, this plays a deterministic role exclusively subject to the laws of nature, and related to the applied sciences. Against this dominance of technology, the role of culture and society would be regarded as simple derivatives. The so-called classical model of technology assessment (Shrader-Frechette 1985; Westrum 1991), focuses primarily on highlighting the negative effects of the implementation and development of technologies. It would be centered in applying a consistent logic in which technology has effects on other (mainly social) systems.

Against the previous deterministic conceptions of technologies, there are also approaches that intend to integrate the implementation and development of technologies in the "social" dimension. This would, for example, be the case of the literature on constructive technology assessment (Wynne 1975). Table 9 presents the main features of this constructive evaluation of technologies, which are then compared with the classic one.

This alternative constructive perspective, more centered on the social dimension, is incorporated in a functionalist logic. This implies that the economic feasibility of the technologies, the development of real markets, efficiency criteria, the effects created by the diffusion of technologies and equity are also considered. Moreover, we believe that technology, in addition to its technical and organizational dimensions, must also be understood as being immersed in values, ideas, and beliefs.

Table 9 Classic and constructive assessment of technologies

Classic assessment of technologies	Constructive evaluation of technologies
Elitist character (centered on scientific knowledge)	Participatory character
Focused on the regulation of the products of technological activity	Focused on the processes by which technologies are created
Reactive approach (impact assessment)	Proactive approach (ex-ante evaluation)
Economic and probabilistic approach	Interdisciplinary and comprehensive approach

Source López Cerezo and Martín Gordillo (2002: 341)

Table 10 Standard and constructivist view on technology

Standard view on technology	Constructivist view on technology
Clear separation between technology and society	The border between society and technology is diffuse
Technological development is autonomous from social and cultural changes	Technology is socially shaped
Technological changes determine sociocultural changes	Technology and society are co-produced
Technology is applied science	The relationship between science and technology is not unidirectional
Technology is neutral	Technology can be inherently political
Technological development follows an internal logic based on technical efficiency criteria and scientific principles	Technological development is the product of diverse cultural forces
Linear model of technological development: design-development-dissemination	Multidirectional model of technological development
Emphasis on the social impacts of technology	Emphasis on processes of social shaping of technology
Clear distinction between the development of technology and its effects	The development and dissemination phases overlap

Source Aibar (2002: 51)

As part of the EED approach, consequences are enriched from the observation of the dynamics occurring in the Internet universe. Contrary to a deterministic view of the evolution of technology, which seeks to minimize its effects on other systems, consequences, as we understand them in our EED approach, are embodied on the dynamics of the Internet ecosystems and the actors that constitute it. That is, consequences are immersed in the actual epigenetic dynamics, and are not simply effects of dominant and a priori determined dynamics. Thus, it can be concluded that our conception of the consequences is closer to constructive technology assessment and social constructivist approaches (Bijker 1995; Aibar 2002).

In this constructivist understanding, culture and technology move hand in hand, although none is contingent on the other. In both approaches, it is necessary to understand the sociotechnical grids hosting technologies, the properties of evolution associated with irreversibility processes, other evolutionary approaches to understand technical change, and of course regulations by the public sector (Table 10).

In our EED approach, we also distinguish between central and peripheral consequences. The former are those generated from the epigenetic dynamics, in our case, of the large Internet business groups. The latter are related to the evolution of the ecosystem itself, but without being in connection with the decisions made by the big Internet business groups. In this section, we have mainly focused on the central consequences, so the analysis of the peripheral ones also remains to be further studied.

References

Adner, R., & Kapoor, R. (2010). Value creation in innovation ecosystems: How the structure of technological interdependence affects firm performance in new technology generations. *Strategic Management Journal, 31,* 306–333.

Ahuja, G., & Katila, R. (2001). Technological acquisitions and the innovation performance of acquiring firms. *Strategic Management Journal, 22,* 197–220.

Aibar, E. (2002). Cultura Tecnológica. In J. M. de Cózar (Ed.), *Tecnología, civilización y barbarie* (pp. 37–62). Barcelona: Anthropos.

Alexa Global Traffic Rank. (2013). http://www.alexa.com/topsites. Accessed September 13, 2013.

App Annie. (2015). App Annie Index: 2014 Retrospective. http://blog.appannie.com/app-annie-index-retrospective-2014/. Accessed November 19, 2015.

Asheim, B., Boschma, R., & Cooke, P. (2011). Constructing regional advantage: Platform policies based on related variety and differentiated knowledge bases. *Regional Studies, 45*(7), 893–904.

Autio, E., & Thomas, L. D. W. (2013). Innovation ecosystems. Implications for innovation management? In M. Dodgson, N. Phillips, & D. M. Gann (Eds.), *The oxford handbook of innovation management* (pp. 204–228). Oxford: Oxford University Press.

Barua, A., Pinnell, J., Shutter, J., & Whinston, A. B. (1999). Measuring the internet economy: An exploratory study. https://www.researchgate.net/publication/228715100_Measuring_the_Internet_economy_An_exploratory_study. Accessed November 19, 2015.

Basole, R. C. (2009). Structural analysis and visualization of ecosystems: a study of mobile device platforms. In Proceedings of the Fifteenth Americas Conference on Information Systems, San Francisco, California, August 6–9, 2009.

Battistella, C., Colucci, K., de Toni, A. F., & Nonino, F. (2013). Methodology of business ecosystems network analysis: A case study in Telecom Italia Future Centre. *Technological Forecasting and Social Change, 80,* 1194–1210.

Bessen, J., & Meurer, M. J. (2008). *Patent failure how Judges, Bureaucrats, and Lawyers put innovators at risk*. Woodstock: Princeton University Press.

Bijker, W. (1995). *Of bycicles, bakelites and bulbs: Towards a theory of sociotechnical change*. Cambridge: MIT Press.

Bijker, W. E., Hughes, T. P., Pinch, T., & Douglas, D. G. (2012). *The social construction of technological systems: New directions in the in the sociology and history of technology*. Boston: MIT Press.

Bingham, C. B., Eisenhardt, K. M., & Furr, N. R. (2007). What makes a process a capability? Heuristics, strategy, and effective capture of opportunities. *Strategic Entrepreneurship Journal, 1,* 27–47.

Bmimatters. (2012). Understanding Twitter business model. http://bmimatters.com/2012/02/18/understanding-twitter-business-model-design/. Accessed November 19, 2015.

Bonde, J. K. (2013). *Valuation of Facebook Inc. 25 July 2013—$30.54*. Copenhaguen Business School.

Bourgeois, L. K., & Eisenhardt, K. M. (1988). Strategic decision process in high velocity environments: Four case studies in the microcomputer industry. *Management Science, 34*(7), 816–835.

Brandenburger, A. M., & Nalebuff, B. J. (1996). *Co-opetition: A revolutionary mindset that combines competition and cooperation in the marketplace*. Boston: Harvard Business School Press.

Breslin, D. (2011). Interpreting futures through the multi-level co-evolution of organizational practices. *Futures, 43,* 1020–1028.

Breslin, D., Romano, D., & Percival, J. (2015). Conceptualizing and modeling multi-level organizational co-evolution. http://www.researchgate.net/publication/271073367_Conceptualizing_and_Modeling_Multi-Level_Organizational_Co-Evolution. Accessed November 19, 2015.

Carrier, M. A. (2012). A roadmap to the smartphone patent wars and FRAND licensing. CPI Antitrust Chronicle. http://ssrn.com/abstract=2050743. Accessed November 19, 2015.

Case Publisher. (2008). Amazon enters the clod computing business. Stanford University, School of Engineering. http://web.stanford.edu/class/ee204/Publications/Amazon-EE353-2008-1.pdf. Accessed November 19, 2015.

Cass, R. A. (2015). Patent litigants, patent quality, and software: Lessons from the smartphone wars. *Minnesota Journal of Law, Science & Technology, 16*(1).

Chesbrough, H. (2003). *Open innovation. The new imperative for creating and profiting from technology*. Boston, MA: Harvard Business School Publishing Corporation.

Chingale, R. (2015). Alice and software patents: Implications for India. *Journal of Intellectual Property Law & Practice, 10*(5), 353–358.

Christiansen, J. K., & Varnes, C. J. (2007). Making decisions on innovation: Meetings or networks? *Creativity and Innovation Management, 16*(3), 282–298.

Clark, K., & Collins, C. (2002). Strategic decision-making in high velocity environments: A theory revisited and a test. In M. A. Hitt, R. Amit, C. E. Lucier, & R. D. Nixon (Eds.), *Creating value: Winners in the new business environment* (pp. 213–239). Oxford: Blackwell Publishing Ltd.

Clarysse, B., Wright, M., Bruneel, J., & Mahajan, A. (2014). Creating value in ecosystems: Crossing the chasm between knowledge and business ecosystems. *Research Policy, 43*, 1164–1176.

Cocotas, A. (2013). *The social media advertising ecosystem explained*. Business Insider, Inc.

Cohen, W. M., & Levinthal, D. A. (1990). Absorptive capacity: A new perspective on learning and innovation. *Administrative Science Quarterly, 35*(1), 128–152.

Corallo, A., Passiante, G., & Prencipe, A. (2007). *The digital business ecosystem*. Cheltenham: Edward Elgar.

Cordes, C. (2006). Darwinism in economics: From analogy to continuity. *Journal of Evolutionary Economics, 16*(5), 529–541.

Corkery, T., Forder, J., Svantesson, D., & Mercuri, E. (2015). Taxes, the Internet and the Digital Economy. *Revenue Law Journal, 23*(1), Article 7.

Cunningham, S. (2011). Mobile patent suits: Graphic of the day. http://blog.thomsonreuters.com/index.php/mobile-patent-suits-graphic-of-the-day/. Accessed November 19, 2015.

Daidj, N. (2011). Media convergence and business ecosystems. *Global Media Journal, 11*(19), 1–12.

Danova. (2014). http://www.businessinsider.com/the-phablet-phenomenon-trends-and-growth-forecast-for-the-device-that-is-taking-over-mobile-sai-2014-6#ixzz3RfnRBifK. Accessed October 19, 2015.

Davis, J. P., Eisenhardt, K. M., & Bingham, C. B. (2009). Optimal structure, market dynamism, and the strategy of simple rules. *Administrative Science Quarterly, 54*, 413–452.

De Agonia, M., Gralla, P., & Raphael, J. R. (2013). Battle of the media ecosystems: Amazon, Apple, Google and Microsoft. *Computerworld*. http://www.computerworld.com/article/2483616/personal-technology/battle-of-the-media-ecosystems--amazon--apple--google-and-microsoft.html. Accessed November 19, 2015.

Dougherty. (2015). http://www.nytimes.com/2015/08/11/technology/google-alphabet-restructuring.html?_r=0. Accessed October 19, 2015.

Dutton, W. H. (2013). Internet Studies: The foundations of a transformative field. In W. H. Dutton (Ed.), *The Oxford handbook of internet studies* (pp. 1–23). Oxford: Oxford University Press.

Eaton, B., Elaluf-Calderwood, S., Sorensen, C., & Yoo, Y. (2011). Dynamic structures of control and generativity in digital ecosystem service innovation: The cases of the Apple and Google Mobile App Stores. http://is2.lse.ac.uk/wp/pdf/wp183.pdf. Accessed November 19, 2015.

Edquist, C., & Zabala-Iturriagagoitia, J.M. (2012). *Innovation system and knowledge-intensive entrepreneurship*: Sweden: CIRCLE Report 2012/03, Lund University.

Eisenhardt, K. M. (1989). Making fast strategic decisions in high velocity environments. *Academy of Management Journal, 32*(3), 543–576.

Eisenhardt, K. M. (2001). Strategy as simple rules. *Harvard Business Review, 79*(1), 107–116.

Eisenhardt, K. M. (2013). Top management teams and the performance of entrepreneurial firms. *Small Business Economics, 40*, 805–816.

Eisenhardt, K. M., & Martin, J. A. (2000). Dynamic capabilities: What are they? *Strategic Management Journal, 21*(10–11), 1105–1121.

Engelen, A., Kube, H., Schmidt, S., & Flatten, T. C. (2014). Entrepreneurial orientation in turbulent environments: The moderating role of absorptive capacity. *Research Policy, 43*, 1353–1369.

Evans, D. S., Hagiu, A., & Schmalensee, R. (2008). *Invisible engines: How software platforms drive innovation and transform industries*. Boston: MIT Press.

Finkle, T.A. (2011) Corporate entrepreneurship and innovation in Silicon Valley: The case of Google, Inc. *Entrepreneurship Theory and Practice, 36*(4), 863–884.

Fischer, T., & Henkel, J. (2012). Patent trolls on markets for technology—An empirical analysis of NPEs' patent acquisitions. *Research Policy, 41*, 1519–1533.

Fransman, M. (2014). *Models of innovation in global ICT firms: The emerging global innovation ecosystems*. Luxembourg: European Union.

Frenken, K., Van Oort, A., & Verburg, T. (2007). Related variety, unrelated variety and regional economic growth. *Regional Studies, 41*(5), 685–697.

Frey. (2015). http://www.futuristspeaker.com/2015/08/future-of-the-internet-8-expanding-dimensions/. Accessed October 19, 2015.

Frosch, R. A., & Gallopoulos, N. E. (1989). Strategies for manufacturing. *Scientific American, 261*(3), 144–152.

Gallego. (2015). http://www.elmundo.es/tecnologia/2015/03/04/54f749e6ca4741470f8b456b.html. Accessed November 19, 2015.

Gartner. (2013). http://www.gartner.com/newsroom/id/2482816. Accessed September 13, 2013.

Gawer, A., & Cusumano, M. (2012). Industry platforms and ecosystem innovation. Presented in DRUID 2012 Conference, Copenhagen (Denmark), June 19–21, 2012.

Geels, F. (2014). Reconceptualising the co-evolution of firms-in-industries and their environments: Developing an inter-disciplinary triple embeddedness framework. *Research Policy, 43*(2), 261–277.

Gómez-Uranga, M., Miguel, J. C., & Zabala-Iturriagagoitia, J. M. (2014). Epigenetic economic dynamics: The evolution of big internet business ecosystems, evidence for patents. *Technovation, 34*(3), 177–189.

Granovetter, M. S. (1973). The strength of weak ties. *The American Journal of Sociology, 78*(6), 1360–1380.

Gruman, G. M. (2012). The dark side of Apple's dominance. *Infoworld*. http://www.infoworld.com/d/mobile-technology/the-dark-side-of-apples-dominance-185538. Accessed November 19, 2015.

Gueguen, G., & Isckia, T. (2011). The borders of mobile handset ecosystems: Is coopetition inevitable? *Telematics and Informatics, 28*, 5–11.

Gueguen, G., Pellegrin-Boucher, E., & Torres, O. (2006). Between cooperation and competition: The benefits of collective strategies within business ecosystems: The example of the software. *Second Workshop on Coopetition Strategy*. http://www.sciencesdegestion.com/travaux/fichierspdf/GGEPBOTcoopetitionmilan.pdf. Accessed 19 November 2015.

Hadida, A. L., Tarvainen, W., & Rose, J. (2015). Organizational improvisation: A consolidating review and framework. *International Journal of Management Reviews* (in press).

Han, W., & Park, Y. (2010). Mapping the relations between technology, product, and service: Case of Apple Inc. In Proceedings of the Industrial Engineering and Engineering Management (IEEM) International Conference (pp. 127–131), Macau, December 7–10, 2010.

Harris, K. D. (2014). Cybersecurity in the Golden State. How California Businesses Can Protect Against and Respond to Malware, Data Breaches and Other Cyberincidents. California Department of Justice, February 2014. https://oag.ca.gov/cybersecurity. Accessed October 19, 2015.

Hernández-Martínez, A. G. (2006). La decisión y su relación con el tiempo: estrategia, procesos e identidad. *Revista Facultad de Ciencias Económicas: Investigación y Reflexión, 14*(1), 23–43.

Herstatt, C., & Kalogerakis, K. (2005). How to use analogies for breakthrough innovations. *International Journal of Innovation and Technology Management, 2*(3), 331–347.

Hobby, P., Harvey, B., & Verma, U. (2014).Cybersecurity and business vitality. What Every Houston-Area Business Leader Needs to Know. Greater Houston Partnership, September 2014.

Holmes. (2015). https://www.linkedin.com/pulse/facebook-vs-twitter-snapchat-who-win-news-wars-ryan-holmes. Accessed August 19, 2015.

Huizingh, E. K. R. E. (2011). Open innovation: State of the art and future perspectives. *Technovation, 31*, 2–9.

Håkansson, H. (1987). *Industrial technological development: A network approach.* London: Croom Helm.

Iansiti, M., & Levien, R. (2004). *The keystone advantage: What the new dynamics of business ecosystems mean for strategy, innovation, and sustainability.* Boston, MA: Harvard Business School Press.

Iansiti, M., & Richards, G. L. (2006). The information technology ecosystem: Structure, health, and performance. *The Antitrust Bulletin, 51*(1), 77–110.

IDC. (2013). http://www.idc.com/getdoc.jsp?containerId=prUS24093213. Accessed September 13, 2013.

Isckia, T., & Lescop, D. (2009). Open innovation within business ecosystems: A tale from amazon.com. *Communications & Strategies, 74*, 37–54.

Iyer, B., & Davenport, T. H. (2008). Reverse engineering Google's innovation machine. *Harvard Business Review*, April, 1–11.

Iyer, B., Lee, C. H., & Venkataraman, N. (2006). Managing in a small world ecosystem: Some lessons from the software sector. *California Management Review, 48*, 28–47.

Jansen, S., Brinkkemper, S., & Cusumano, M. A. (2013). *Software ecosystems: Analyzing and managing business networks in the software Industry.* Cheltenham: Edward Elgar.

Jing, Z., & Xiong-Jian, L. (2011). Business ecosystem strategies of mobile network operators in the 3G era: The case of China Mobile. *Telecommunications Policy, 35*, 156–171.

Judge, W. Q., & Zeithaml, C. P. (1992). Institutional and strategic choice perspectives on board involvement in the strategic decision process. *Academy of Management Journal, 35*(4), 766–794.

Karakas, F. (2009). Welcome to World 2.0: The new digital ecosystem. *Journal of Business Strategy, 30*(4), 23–30.

Kay, L., Newman, N., Youtie, J., Porter, A. L., & Rafols, I. (2014). Patent overlay mapping: Visualizing technological distance. *Journal of the Association for Information Science and Technology, 65*(12), 2432–2443.

Knable Gotts, I., & Sher, S. (2012). The particular antitrust concerns with patent acquisitions. *Competition Law International*, August, 30–38.

Lehman, G., & Haslam, C. (2013). Accounting for the Apple Inc business model: Corporate value capture and dysfunctional economic and social consequences. *Accounting Forum, 37*, 245–248.

Levy, S. (2015). How Twitter found its money mojo. https://medium.com/backchannel/how-twitter-found-its-money-mojo-1d170e3df985. Accessed November 19, 2015.

Leyden, A., & Dolmans, M. (2014). The Google Commitments: Now with a Cherry on Top. *Journal of European Competition Law & Practice, 5*(5), 253–255.

Leydesdorff, L., Kushnir, D., & Rafols, I. (2014). Interactive overlay maps for US patent (USPTO) data based on International Patent Classification (IPC). *Scientometrics, 98*(3), 1583–1599.

Li, Y. R. (2009). The technological roadmap of Cisco's business ecosystem. *Technovation, 29*, 379–386.

Lim, D. (2014). Standard essential patents, trolls and the smartphone wars: triangulating the end game. http://ssrn.com/abstract=2495547. Accessed November 19, 2015.

López Cerezo, J. A., & Martín Gordillo, M. (2002). Evalución de tecnologías en contexto social. In J. M. de Cózar (Ed.), *Tecnología, civilización y barbarie* (pp. 336–358). Barcelona: Anthropos.

Lundvall, B.-Å. (1992). *National innovation systems: Towards a theory of innovation and interactive learning.* London: Pinter Publishers.

Malik. (2015). http://www.newyorker.com/tech/elements/in-silicon-valley-now-its-almost-always-winner-takes-all. Accessed December 30, 2015.

Manjoo, F. (2011). The great tech war of 2012. http://www.fastcompany.com/1784824/great-tech-war-2012. Accessed November 19, 2015.

March, J. G. (1991). Exploration and exploitation in organization learning. *Organization Science, 2*, 71–87.

Moore, J. F. (1996). *The Death of competition: Leadership and strategy in the age of business ecosystems*. New York: Harper Business.

Moore, J. F. (1998). The rise of a new corporate form. *Washington Quarterly, 21*(1), 167–181.

Moore, J. F. (2005). Business ecosystems and the view from the firm. *The Antitrust Bulletin.* http://cyber.law.harvard.edu/blogs/gems/jim/MooreBusinessecosystemsandth.pdf. Accessed November 19, 2015.

Mortara, L., & Minshall, T. (2011). How do large multinational companies implement open innovation? *Technovation, 31*(10–11), 586–597.

Nachira, F. (2002). Towards a network of digital business ecosystems. Discussion Paper. ICT for Enterprise Networking, European Commision. http://ec.europa.eu/information_society/topics/ebusiness/godigital/sme_research/doc/dbe_discussionpaper.pdf. Accessed November 19, 2015.

Nazzini, R. (2015). Google and the (Ever-stretching) Boundaries of Article 102. *Journal of European Competition Law & Practice, 6*(5), 301–314.

Ndou, V., Schina, L., Passiante, G., Del Vecchio, P., De Maggio, M. (2010). Toward an open network business approach. In Proceedings of the 4th IEEE International Conference on Digital Ecosystems and Technologies (pp. 282–287), Dubai, April 13–16, 2010.

Nielson, S (2014). Why Apple's ecosystem is its biggest competitive advantage. http://marketrealist.com/2014/02/ecosystem/. Accessed November 19, 2015.

OECD. (2013). *The app economy*. Directorate for science, technology and industry. Committee for information, computer and communications policy. Paris: OECD.

Paik, Y., & Zhu, F. (2013). The impact of patent wars on firm strategy: evidence from the global smartphone market. http://ssrn.com/abstract=2340899. Accessed November 19, 2015.

Palmié, M., Lingens, B., & Gassmann, O. (2015). Towards an attention-based view of technology decisions. *R&D Management* (in press).

Parrilli, M. D., & Zabala-Iturriagagoitia, J. M. (2014). Interrelated diversification and internationalisation: Critical drives of global industries. *Revue d'Économie Industrielle, 145*, 63–93.

Pellegrin-Boucher, E., & Gueguen, G. (2004). How to manage co-operative and coopetitive strategies within IT business ecosystems, the case of SAP, the ERP leader. In EIASM Workshop on coopetition strategy: Toward a new kind of interfirm dynamics? September 16–17, 2004, Catania (Italy).

Peltoniemi, M., & Vuori, E. (2004). Business ecosystem as the new approach to complex adaptive business environments. In Proceedings of eBusiness research forum (pp. 267–281). http://citeseerx.ist.psu.edu/viewdoc/download?doi=10.1.1.103.6584&rep=rep1&type=pdf. Accessed November 19, 2015.

Razavi, A. R., Krause, P. J., & Strømmen-Bakhtiar, A. (2010). From business ecosystems towards digital business ecosystems. In Proceedings of the 4th IEEE International Conference on Digital Ecosystems and Technologies (pp. 290–295), Dubai, April 13–16, 2010.

Risku, J. (2012). Ecosystems: Apple, Google, Microsoft. https://abstractionshift.wordpress.com/2012/02/19/mobile-internet-ecosystems-apple-google-microsoft/. Accessed November 19, 2015.

Rong, K., Hu, G., Lin, Y., Shi, Y., Guo, L. (2015). Understanding business ecosystem using a 6C framework in internet-of-things-based sectors. *International Journal of Production Economics, 159*, 41–55.

Rothschild, M. (1995). *Bionomics economy as ecosystem*. New York: Henry Holt and Company.

SEC. (2013). United States Securities and Exchange Commission. Washington, DC 20549. Form 10-K for the fiscal year ending December 31, 2013. Commission File Number 001-36164. Twitter, Inc. http://www.sec.gov/Archives/edgar/data/1418091/000095012314003031/twtr-10k-20131231.htm. Accessed August 15, 2015.

Shearman, S, (2012). http://www.marketingmagazine.co.uk/news/1136805/Amazon-set-launch-iTunes-rival/. Accessed October 19, 2015.

Shrader-Frechette, K. (1985). Technology Assessment, Expert Disagreement, and Democratic Procedures. Research in Philosophy & Technology (Vol. 8). JAI Press: New York.

Skelly, P. (2014). The new business operating system: combining office 365 and the Microsoft cloud ecosystem to create business value presentation. http://www.threewill.com/new-business-operating-system-combining-office-365-microsoft-cloud-ecosystem-create-business-value-presentation/. Accessed November 19, 2015.

Spencer, G. (2015). No ecosystem is an island: Google, Microsoft, Facebook & Adobe's iOS Apps. http://www.macstories.net/stories/no-ecosystem-is-an-island-google-microsoft-facebook-adobes-ios-apps/. Accessed November 19, 2015.

Statista. (2015). The statistics portal. http://www.statista.com/. Accessed November 19, 2015.

Tardiff, T. J. (2015). Net neutrality: Economic evaluation of market developments. *Journal of Competition Law & Economics* (in press).

Teece, D. J. (2007). Explicating dynamic capabilities: The nature and microfoundations of (sustainable) enterprise performance. *Strategic Management Journal, 28*, 1319–1350.

Thomas, L. D. W., & Autio, E. (2012). Modeling the ecosystem: a meta-synthesis of ecosystem and related literatures. Presented in DRUID 2012 Conference, Copenhagen (Denmark), June 19–21, 2012.

Trewe, M. (2011). Platlas for Facebook: Social network ecosystem exploration tool. http://theamericangenius.com/social-media/platlas-for-facebook-the-first-social-network-ecosystem-exploration-tool/. Accessed November 19, 2015.

Useche, D. (2015). Patenting behaviour and the survival of newly listed European software firms. *Industry and Innovation, 22*, 37–58.

Van de Brande, V., de Jong, J. P. G., Vanhaverbeke, W., & de Rochemont, M. (2009). Open innovation in SMEs: Trends, motives and management challenges. *Technovation, 29*, 423–437.

Van Eck, N. J., Waltman, L., van Raan, A. F. J., Klautz, R. J. M., & Peul, W. C. (2013). Citation analysis may severely underestimate the impact of clinical research as compared to basic research. *PLoS ONE, 8*(4), 1–6.

Van Zeebroeck, N., Pottelsberghe, Van, de la Potterie, B., & Guellec, D. (2009). Claiming more: The increased voluminosity of patent applications and its determinants. *Research Policy, 38*(6), 1006–1020.

Wan, J., Zhan, H., Wan, X., & Luo, W. (2011). The business ecosystem of the Chinese software industry. *iBusiness, 3*(2),123–129.

Westrum, R. (1991). *Technology & Society: The shaping of people and things.* Belmont: Wadsworth.

Wiethaus, L. (2015). Google's favouring of own services: Comments from an economic perspective. *Journal of European Competition Law & Practice, 6*(7), 506–512.

Wirtz, B. W., Mathieu, A., & Schilke, O. (2007). Strategy in high-velocity environments. *Long Range Planning, 40*(3), 295–313.

Wynne, B. (1975). The rhetoric of consensus politics: A critical view of technology assessment. *Research Policy, 4*, 108–158.

Zahra, S. A., & George, G. (2002). Absorptive capacity: A review, reconceptualization, and extension. *Academy of Management Review, 27*(2), 185–203.

GAFAnomy (Google, Amazon, Facebook and Apple): The Big Four and the b-Ecosystem

Juan Carlos Miguel and Miguel Ángel Casado

1 Introduction

Information and Communication Technologies have fostered the emergence of companies such as Google, Amazon, Facebook and Apple, known as GAFA, on the Internet. These companies are interrelated, form the business ecosystems (b-ecosystems) and have a core activity which they complement with others.

The term GAFA was first used in 2011 by Simon Andrews, chief of Addictive Marketing agency in London, with the aim of studying the marketing strategies of these companies and their applicability to other sectors. The term was also used in France in December 2012 (Ducourtieux 2014) referring to the big Internet companies that are tax-exempt in France. The Financial Times refers to GAFA (Rachman 2015), although the purpose is to underline their industrial and even cultural dominance (Chibber 2014). In this text, GAFA is used to refer four companies that share certain features (they are two-sided markets which compete as ecosystems, their DNA contains innovation and they are quasi-monopolies in their core activities).

The aim of this chapter is to show which factors contribute to explaining their quasi-monopolistic position and their influence on the development of the markets

J.C. Miguel (✉) · M.Á. Casado
Audiovisual Communication and Advertising Department,
University of the Basque Country, UPV/EHU, Bilbao, Spain
e-mail: jc.miguel@ehu.eus

M.Á. Casado
e-mail: miguelangel.casado@ehu.eus

© Springer International Publishing Switzerland 2016
M. Gómez-Uranga et al. (eds.), *Dynamics of Big Internet Industry Groups and Future Trends*, DOI 10.1007/978-3-319-31147-0_4

where they operate. The mentioned position confers them offensive and defensive power as it is only possible to compete with them ecosystemically, or, in other words, with a similar group of related activities.

Some of the features of GAFA are analysed in this study; particularly the competition between them in spite of their being extremely different. The intention is to explain the growth and size of these firms via their characterisation as two-sided markets. However, they not only compete with each other but with all firms that offer advertising (this applies to Facebook and Google) and those that sell devices and content (Apple) and even online and offline shops that sell Amazon products. This competition places pressure on the pace of innovation and firms are constantly obliged to bring out new and improved products or services, some of which are surprising. Nevertheless, part of the innovation focuses on exploring new financing sources which compensate the limitations of advertising—Facebook and Google—or show significant growth potential, which exceeds that which characterises the core activity—Amazon and Apple. Contrary to appearances, this innovation does not add extreme diversification but is integrated in each firm's ecosystem, which it strengthens.

First, the literature on business ecosystems is reviewed, as this term can be used to characterise each GAFA member and we can therefore make use of its metaphorical power. After proposing a definition applicable to the scope of study for the firms considered in this research, some characteristics of ecosystems are put forth, various aspects of this special competition between them are analysed, in addition to innovation, which raises new and interesting questions. Each GAFA group is the leader in its core activity (e.g. Google controls 90 % of all web searches, Facebook has a 75 % share of the social networks, Amazon 6 % of global online sales and Apple 45 % of smartphone web traffic) (Fabernovel 2013). This study does not include Twitter, Microsoft, Yahoo or Samsung, which may also be studied as business ecosystems and could be considered an extension of GAFA (see Chapter "Epigenetic Economics Dynamicsin the Internet Ecosystem" in this book by Zabala-Iturriagagoitia et al.). The four GAFA members suffice for the study of ecosystem dynamics, particularly the competition between them. Nevertheless, practically all of the points raised in this research can be applied to firms expanding from GAFA.

Their business models and ways of obtaining revenues differ. Whereas, Facebook and Google obtain almost all their revenues from advertising; Amazon and Apple only obtain marginal amounts from this source. Although the degrees of internationalisation are different, they are important.

Apple's market cap of 8,000,000 dollars per employee is outstanding. All the companies except Amazon exceed Microsoft's revenues per employee, which was 737,992 dollars in 2014. If it is compared with Walmart, which has 2.2 million employees, Amazon nearly triples its revenue per employee. However, Walmart has higher revenues than GAFA ($473,080 million—Statista 2014a). Sales per employee ratio enable us to predict high profitability. However, Amazon has negative profitability.

2 Dynamics of Business Ecosystems

Each period may be characterised by the words and concepts generally used, first appearing in the scientific or academic field and later included in all types of texts. Concepts such as globalisation, localisation, convergence, business models, digital business ecosystems, innovation ecosystems or business ecosystems have become widely used and are found in all types of texts—academic, professional, blogs…—and contexts. They therefore lose their original meaning, which makes it necessary to go back over them from time to time.

This is particularly necessary in the case of the term *ecosystem*, which covers a very broadly defined concept and is also often used to replace business ecosystems. In effect, according to Kelly (2015), *"Ecosystems are dynamic and co-evolving communities of diverse actors who create and capture new value through increasingly sophisticated models of both collaboration and competition"*. This definition makes it possible to refer to regional innovation systems, to a specific industry (media ecosystem) to telecommunications, or mobile payment systems, etc. Rather than encompassing only the relationships found in the value chain, business ecosystems go beyond them as they involve various firms from other industries (Rong et al. 2013a, b). Rong et al. (2015) apply the concept of business ecosystem to the Internet of Things (IoT) and highlight that IoT should be seen more like an ecosystem than a supply chain. This study focuses on explaining the new situation which occurs as industries (hardware, networks and content) interact, and also inside such industries.

Based on the existing literature, Pilinkienė and Mačiulis (2014) defined five typologies of ecosystems (industrial ecosystem, innovation ecosystem, business ecosystem, digital business ecosystem and entrepreneurship ecosystem). Their analysis took into account the actors present, the environment, the micro and macro impacts and the key determinants affecting the ecosystem.

A striking feature of business ecosystems is that the actors are "customers, competitors, market intermediaries, companies selling complementary products and suppliers" (p. 367), in addition to the "regulatory agencies and media outlets that can have a less immediate, but just as powerful, effect on your business" (p. 367). Then, competitors and customers also form part of b-ecosystems.

After analysing the tables proposed by Pilinkiene and Mačiulis (2014), it can be stated that business ecosystems and digital business ecosystems are very similar. However, the latter first appeared in European Union documents and looked to underscore that the convergence of networks and information technologies form the basis of the economy's growth and development, which is, in other words, use of policy in the industry (World Economic Forum 2015). This intervention by government institutions is also found in the entrepreneurship system. The presence of the public sector has also been examined by Clarysse et al. (2014), who demonstrate the role that public funding has played in a certain region of Flanders. For this reason, the intervention of public agencies and administrations is currently thought to be fundamental when forming an emerging industries ecosystem (Moritz 2015).

Mars et al. (2012) made a noteworthy contribution to marketing theory with their studies of the similarities and differences between biology ecosystems and organisational ecosystems. The stability of an organisation and of a biology ecosystem both depend on the keystone actors and both give rise to mutualism, commensalism and predation. Furthermore, nestedness infers resiliency, so the disappearance of an actor in a nested network does not imply immediate collapse. They point out many differences and mainly base their view on the fact that there are human beings in organisations. Humans tend to make predictions about the future and later set objectives and strategies to reach them. In addition, alternative plans are always considered in case the original one fails. Their most valuable idea is that business ecosystem is an interesting metaphor which should be maintained, above all to study the creation and/or development of new ecosystems.

2.1 GAFA as Ecosystems

According to Moore (1996), the contribution of the new concept was that it went beyond the concept of industries. In the case of GAFA, our focus is closer to Moore's definition. Each GAFA member (Google, Amazon, Facebook and Apple) may be considered a business ecosystem because each one's group of activities belong to different "industries", and such activities as a whole are related and used by millions of people.

What is of interest concerning this point is to keep the interconnection between the activities of the different industries, which converge inside companies. The definition of business ecosystem proposed in this study therefore captures the spirit of Moore (1998, 2015), the forerunner of business ecosystem, and is inspired by the definition given by Makinen and Dedehayir (2012). Business ecosystem is the interrelation between activities, which, as a whole, form an integrated system of hardware, software and content in variable proportions. This system is managed by a firm or business group and offers consumers higher value than if they accessed the different activities provided by various firms separately. In this way, the activities inside firms are considered without overlooking the context in which such firms operate. This is formed by all the firms found in the broad hardware, telecommunications or content industry. The underlying idea of the definition is that the concept of business ecosystem can be applied to the whole of the hardware, software, networks and content, as Fransman does (2014), and also to each GAFA group's activities as a whole. Twitter, Microsoft or Samsung may also be approached as business ecosystems, although GAFA have some special characteristics that require study.

The concept of business ecosystem makes it possible to understand activities which, in some cases, might seem very different and unrelated, and that there is a symbiotic relationship between them, preventing the consumer, in practice, from doing without the rest of the activities. It scarcely makes sense for people to buy an iPhone and not use any applications, or any content on iTunes. Consumers find more advantages by using all the options that the ecosystems provide. It is therefore interesting to note Amazon's recommendations, which are sent after time has elapsed, and can quickly result in a purchase. Another example is the Amazon Prime service, which ensures free delivery for a small fee and allows free access to video and music content.

Rather than sell devices (hardware), software, content and applications separately, Apple sells the ecosystem as a whole. Customers enter the system via the hardware (which has the integrated software), and the content and applications form part of the ecosystem since the hardware is not bought for itself alone. Entrance and exit from the ecosystem carry costs that go beyond the first purchase (Borrow 2014). Once the customer enters the ecosystem, exit involves costs that may be significant because the applications and content are not easily transferred to devices from other ecosystems such as Samsung. In the case of Apple, it consists of moving to other similar devices, although the ergonomics are different and content is lost. When a customer leaves Facebook, it may be impossible to delete the information that has been posted. Approaching ecosystems as a group of related items means that the firm's medium and long-term innovation and management must envisage them as a whole. Companies may be the first to develop a new item such as the smartwatch, but they should also consider the applications and uses it will be given.

The first consequence for firms is that they are going to compete ecosystemically, or through a group of activities rather than device by device or application by application. This means that only a few can continue growing and innovating since they need an ecosystem to do so. It should be noted that Nokia's loss of market share was the result of the firm's failure to build an ecosystem that could compete with other big ones.

Ecosystems never stop advancing, growing and creating continuous sources of instability. As regards applications downloaded on smartphones, Apple's or Google's ecosystem, for instance, is strengthened by Android but this also involves introducing a new source of possible problems. Application developers are outside and do not belong to the group of ecosystem firms. This feature of growth and complexation is known as "emergence" in biology (Matthies et al. 2015).

3 Innovation in GAFA

They devote extraordinary amounts of resources to innovation. In absolute figures, Google tops the list of investment in innovation. Its Google X division created the driverless car; Google Glasses, the vibrating spoon for Parkinson's sufferers; the contact lens that monitors blood glucose, etc. The percentages are extremely high and predicted to grow or continue at their present levels, as a result of the competitive pressure and pace of innovation the company is distinctive for. Table 1

Table 1 GAFA. Some financial information. In dollars, units (employees) and % (RD/sales)

	Facebook	Apple	Amazon	Google
Market cap	$225,840,000	$741,850,000	$198,380,000	$369,110,000
Sales	$14,470,000	$182,350,000	$88,990,000	$65,830,000
Revenues	Advertising 93 % Other 7 %	56 % iPhone	25 % Media 75 % Electronic and other, including AWS	89.5 % advertising 10.5 % other
Revenues outside the US	>50 %	70 %	38 %	57 %
Employees	9199	92,600	154,100	53,600
Market cap/Sales	15.60	4.06	2.3	5.60
Revenue per employee	$1,574,537	$1,969,222	$577,482	$1,228,171
Net Margin	23.46 %	21.67 %	−0.27 %	21.16 %
Research and Development	$2,666,000	$6,041,000	$9,275,000	$9,830,000
R&D/sales	18.4 %	3 %	9.7 %	15 %
R&D (adjusted)		10.4 %		
Net revenue	$2,930,000	$6,040,000	($241,000 M)	$13,930,000
Users accounts	1415 million monthly active users in 2015 (Statista 2014b) – In January 2015, Whatsapp had 700 million users who sent 30,000 million messages per day (Kokalitcheva 2015)	800 million iTunes account holders in late 2013, most (but not all) of whom have credit cards on file (Marous 2014)	270 million active customers. In 2014 Amazon 'only' had 41.3 million unique buyers in the United States (Marous 2014)	Google+300 million monthly active users in October of 2013 (Marous 2014) 49,000 searches per second (Internet Live Stats 2015)

Source MarketWatch and Bloomberg. The market cap data are from 18-5-2015. Sales are shown on Form 10-K. Research and Development data are from Google Finance, obtained from Form 10-K. FB ends the fiscal year on 31-12-2014. Apple on 27-9-2014, Amazon, on 31-12-2014 and Google on 31-3-2015

highlights the relatively low percentage that Apple invests in innovation. This can be explained by the innovation system that the firm has in place and which is worthy of study.

However, its innovation efforts should be clarified. High percentages are funnelled into innovation although acquisitions must also be taken into account as many of them contribute to innovation.

The study of its acquisitions enables us to identify what the group's keystone activities are. However, above all, it allows for a posteriori observation of what its strategic concerns were. It is quite easy to analyse in the case of Apple, as many of its activities are hardware- and software-related. In 2014, it acquired four map-related companies and in 2013, two search-related companies. Maps were not

Table 2 GAFA's investments in acquisitions from 2011 to 2014 ($Millions)

		2011	2012	2013	2014
Google	Properties and equipment	3.438	3.273	7.358	10.959
	Intangibles	1.900	10.568	1448	4.888
	Total	5.338	13.841	8.806	15.847
Facebook	Properties and equipment	606	1.235	1.362	1.831
	Business and intangibles	24	911	368	4.975
	Total	630	2.146	1730	6.806
Apple	Business acquisition	244	350	496	3.765
	Properties, plants and equipment	4260	8.295	8.185	9.571
	Intangibles	3.192	1107	911	242
	Total	7.696	9.752	9.592	13.578
Amazon	Properties and equipment, including internal software and website development	1.811	3.785	3.444	4.983
	Acquisitions	705	745	312	979
	Total	1.516	4.530	3.756	5.962

Source United States Securities and Exchange Commission, Form 10-K. Google ends the fiscal year on 31-12, Facebook and Amazon on 31-1, and Apple on 27-9

Apple's strongest point, and its awareness of mobiles using an increasing amount of applications with geolocalisation and maps prompted it to buy companies with that technology. Google has bought several companies to integrate them in Google X, which is developing the driverless car, the contact lens that monitors blood glucose, etc. Each GAFA member has its own style. To date, Apple has not bought any company for over 400 million dollars (Dilger 2014). Facebook has made the largest purchase so far with 19,000 million for WhatsApp, in an effort to reinforce its core activity, which is a social network platform.

Acquisitions should also be seen as complementary to innovation. In effect, acquisitions are one of the ways to obtain innovations, and exceptional human resources at the same time because the owners are not only excellent technicians (engineers, computer scientists) but also entrepreneurs. Acquisitions therefore mean innovative talent entering the firm. Table 2 shows GAFA's acquisitions. Each of the firms itemises its acquisitions differently and it is difficult to determine the exact figure showing acquisitions that may be identified as innovation. Even if only half of the acquisitions registered correspond to innovation, Google's innovation effort would be 25 %, Amazon's 13 %, Apple's 6.7 % and Facebook's would exceed 30 % of its sales figures.

The extraordinarily fast pace of innovation can become intolerable and difficult for firms to sustain. Not only do they design new objects or services (watches, cars, glasses) but are required to include increasingly innovative features in existing devices. This means that firms must strike a balance between innovation costs and their profitability, requiring that they introduce logic which is not always popular with consumers. On the foreign scenario, the patent wars between various telephone companies, above all in the United States, are hindering innovation itself. A mechanism (patents) which, in theory, serves to shore up innovations, involves high legal costs due to the numerous lawsuits they are drawn into. If we consider a smartphone, it is impacted by over 250,000 patents, the staggering difficulty of complying with them all is obvious.

4 Ecosystemic Competition. GAFA Against Each Other

Item 1A-Risk factors—in the 10-K form, which all publicly traded companies must file with the United States Securities and Exchange Commission, show the enormous competition all GAFA face. Although competition goes beyond GAFA, this study focuses on that which occurs inside them.

Table 3 GAFA. Competitors according to item 1A of the 10-K (risk factors)

Amazon	Google	Apple	Facebook
1. Physical-world retailers, publishers, (…) of our products 2. Other online e-commerce and mobile e-commerce sites 3. Media companies, web portals, comparison shopping websites, web search engines, and social networks 4. Companies that provide e-commerce services, including website development, fulfilment, customer service, and payment processing; companies that provide information storage or computing services 5. Companies that design, manufacture, market, or sell consumer electronics, telecommunication, and electronic devices	1. General purpose search engines 2. Vertical search engines and e-commerce websites (Kayak, Linkedin, Amazon, etc.) 3. "Social networks, such as Facebook and Twitter" 4. Other forms of advertising, (television, radio, …) 5. Other online advertising platforms like Facebook, that compete for advertisers with AdWords 6. Other operating systems and mobile device companies 7. Providers of online products and services	1. Hardware 2. Other digital electronic devices 3. Software 4. Online services and distribution of digital content and applications	1. Companies that replicate the range of communications and related capabilities we provide, like Google+Mobile applications, that provide communications functionality (messaging, photo- and video-sharing,…) web—and mobile-based information and entertainment products and services that are designed to engage people and capture time spent online and on mobile devices. Traditional, online, and mobile businesses providing media for marketers to reach their audiences

Note The texts are literal, as the companies appear in the 10-K reports, although they have been summarised except in the case of Apple. All the lists are from 2014 although the dates of their fiscal years differ

Competitors are one of the points specified in the 10-K form, although each company itemises them differently. When studying Table 3 on Competitors, Amazon, Google and Facebook appear in addition to the others which are explicitly shown. In Amazon's disclosure of competitors, 3 and 5 refer to Google. In Google's list, 2 refers to Amazon and 3 to Facebook. Facebook lists Google as 1 and Apple and Amazon as 3.

Amazon competes with all types of online and offline shops that sell similar products and devices, which demonstrates that its scope of competition includes firms outside GAFA. However, upon studying this group of firms, it is observed that they all operate in the others' fields, with varying degrees of intensity. The fact that all of them are active in nearly every one these fields make us wonder why they behave in this manner. The first reason is general: they are all aware that they move in a wider ecosystem which is formed by materials, networks and content (Fransman 2014). Furthermore, they see that Apple is benefiting from the ecosystem which it has built based on devices, content and applications and that the new areas have growth potential. They have all seen that there is a tendency to use mobiles for all types of activities, and not only those related to communication, relationships or entertainment, so they are going to do everything possible to gain a stronger foothold. Google has developed software for mobiles and has even launched manufacture of devices while creating an application platform to safeguard its leading position as a search platform via mobiles.

By offering the same products and services as the others, they are not seeking confrontation in these fields but look to place obstacles in the leaders' paths. Google+ could not replace Facebook, but seeks a bigger market share to make it harder for Facebook to grow.

GAFA are hegemonic in their core businesses, understood as those which are shown in Table 4. Haucap and Heimeshoff (2014) point out that "Google, YouTube, Facebook and Skype are typical examples of Internet firms that currently dominate their relevant markets and who leave only limited space for a relatively small competitive fringe". The same could be said about GAFA, although with some nuances. First, not all the firms that form GAFA create content— except Google's YouTube— although GAFA are aware of its importance; and this is expected to change as Amazon Original Movies has already shown interest in audiovisual production.

In his Mac Taggart lecture at the International Television Festival in August 2011 in Edinburgh, Google CEO Eric Schmidt defended his company against accusations of copyright infringement and use of other owners' content (Schmidt 2011). The journalist J. Robinson made a good summary: "good content drives search, and search drives advertising. The more compelling the content there is online, the more money Google makes" (Robinson 2011). Therefore, relationships with content owners are essential for GAFA, especially for Google and Facebook due to their advertising-based funding model, and which seems to have been inspired by media funding models, especially radio and general television, which is ad financed. If content is consumed via mobile devices such as smartphones

Table 4 Competition in GAFA, by business activities

	Google (G)	Amazon (Am)	Facebook (F)	Apple (Ap)
Definition	Advertising	Commerce	Social networks	Hardware/software/applications high-end consumer devices at premium prices
Competitors	Searching	Giant virtual mall		
	A tiny profit from each of billions of online ad sales			
Search engines	Google search (90 % share in search engines)	Its search engine is used for products and books, 6 % of online retail (Fabernovel)	"Testing an in-app keyword search engine that lets you find websites and articles to add to your status updates" 75 % share of social media	*Siri*
	Thematic browsers (Kayak)			*Applebot*
				45 % share of Smartphone web traffic. It could compete with Google in future
Retail	Google same day	Amazon fresh	Facebook "Buy" button	Apple Stores online
	Non-perishable products	Amazon.com		
Physical retail	One shop in London in 2015	1 physical store in Manhattan		Apple Stores
Telecom	Mobile plans with Tmobile and Sprint	Its web capacity enables its activity as a telecom operator		Relations with telecommunications companies to offer their mobiles
Health	Google Fit (app)		Interested in developing	Apple Health (app)
	Google X			
Energy	Nest Labs			The only one that uses renewable energy in all its stores
	Investment in solar energy with Atlantic Grid Development to provide electricity in New Jersey			It has several solar farms
				Invests in electricity in California
GFinance App	Google Wallet	Leaves mobile-wallet (app) in 2015	Friend to friends payments	Apple Pay

(continued)

Table 4 (continued)

	Google (G)	Amazon (Am)	Facebook (F)	Apple (Ap)
Car	Google Car Android Car Platform			Car Play Platform iCar Projects
Drones	Titan Aerospace	Ascenta		iDrones
Apps	Android Apps store	Amazon Apps store		Apple Apps store
Wearables	Google Watch		Oculus	iWatch
News	Google News		Instant articles	Apple Apps news for news
Television	Google Play Google Chrome	Amazon FireTV		Apple TV
Hardware	Nexus	Echo (voice recognition) Kindle Fire Phone	Smartphone with HTC	iPhone iPad Mac iPod iOS
Software	Android			
Music and video	Google Music/YouTube	Amazon MP3 Store Amazon Instant video	Deals with Spotify and Netflix	iTunes Streaming (beats)
News	Google News		Instant articles	Through apps from Apple Store
Maps	Google Maps			Apple maps
Books	Google Books	Amazon begun with books	Books on Facebooks	iBooks
Telecom & IT	Google Fiber	Cloud Drive	WhatsApp	Apple SIM
Health	The Life Sciences division of Google X		Everyday Health	Apple's Health kit
Cloud	Google Drive	Amazon Cloud Drive Amazon Web services	De facto, photo sharing	iCloud Drive

Sources Own elaboration based on Elmer-DeWitt 2014; Fabernovel, Constine 2015, Efrati 2013; Subramanian 2014; Martin 2014; Poulter 2014; Mannes 2014; Pereira 2014; Somaiya et al. 2015; Bensinger 2013; Saviolo 2014 and MarketWatch 2014.

and tablets, firms have to move into the hardware and software industries and applications.

Each group has a rival or fierce competitor either inside or outside GAFA. The biggest rivalry inside GAFA is between Google and Facebook, as both depend on online advertising (Jhonsa 2015). However, the rivalry between Google and Apple is also remarkable. Outside GAFA, there is the case of Apple, with Samsung as its rival. They compete on the smartphone and tablets market and are also engaged in furious patents competition. Apple has innovated and continues to do so whereas Samsung is reaching a market share that enables it to overshadow Apple. Amazon competes with all businesses that sell products online and offline.

5 Ecosystems and Platforms

GAFA are often called platforms (Perrot 2011) because different types of agents converge in them: content producers, advertisers, sellers, application developers, etc., and platform is identified with business model (Simon 2011). Some authors identify platform with business ecosystem (Travlos 2013).

GAFA are intermediaries, which can be defined as platforms. They act as intermediaries between news search results (Google); between individuals and between news, as visits to news sites are increasingly initiated on Facebook (Rebillard 2010), between book publishers and buyers (Amazon), between buyers and music editors (iTunes). As these are global firms, this intermediation gives them enormous power in negotiations with content producers, meaning that they can impose rules (certain prices on books, the percentage charged by developers, free access to news headlines).

These are platforms characterised by two-sided markets (Eisenmann et al. 2006). In their core activities, Google, Facebook and Amazon are distinctive for being intermediaries between different types of agents. Devices may be seen as platforms in Apple because the users are on one side and the applications on the other. iTunes is situated between content providers and buyers.

Markets are classified as two-sided if indirect network effects exist (Rysman 2009). They exist if growth in the number of users on one market attracts those from the other. The increase in the number of Facebook users attracts more advertisers to use Facebook (Table 5). By definition, the presence of indirect effects implies that the bigger one group of users is, the bigger the other one will be, which means that size is directly related to exploitation of indirect network economics and, as a result, high concentration levels may be found. Netflix, Spotify and GAFA fulfil the size requirement, or, more specifically, some features that such size can foster. Thus, a site offering music must offer the largest amount or the widest variety, or even better, both.

GAFA are actually multi-platforms because they are multi-sided. Google places users, who search for goods and services, on one side and advertisers on the other. In this instance, it acts as intermediary between content and consumers. However, the content is heterogeneous—news, books, music, videos—so, it could be argued that *two-sided markets* exist for each of the activities. One would be Youtube, for Google; Google Books would be another for books and Google News for news. Each line of business in the competitive matrix can be considered a two-sided market. A further characteristic is that they are constantly expanding their range of activities, which multiplies the number of platforms. Moreover, it could even be considered a different market from mobile telephony because it raises different challenges for each of the activities.

There are doubtlessly factors which limit the growth of these platforms in which indirect network effects occur. Haucap and Heimeshoff (2013) argue that "capacity limits, product differentiation and the potential for multi-homing exercise negative effects on market concentration". GAFA grow in a modular way. In other words, growth can take place by adding extensions so as not to have to rethink the entire ecosystem. Google and Facebook have designed their own search and storage software so that computers can be added without having to constantly remodel the entire group of activities. A look at Amazon's history shows how it has increased its business scope; it started out with books and now offers all types of content (Lawson 2014). It is difficult to explain the interconnections between the 900,000 or the 400,000 computers that Google and Amazon use, respectively, for storage, and have to continuously increase to make room for their new activities (Storage Servers, 2013).

6 Opening up Ecosystems. Samsung/Apple

Keeping an ecosystem closed has its pros and cons. The advantage is that everything in the ecosystem may have something in common. For instance, in Apple it is well-designed premium products (Lessin 2013). It seeks to keep its systems closed for the time being, with new applications being included in the system. These new additions sometimes open up new lines of business with enormous potential revenues. This is the case of mobile payment systems. Although Facebook has a payment and transfer system for small amounts of money between users, and Amazon is interested in participating, it is mainly Apple and Google that have been competing for several years. Each of them is looking for an application to include in the closed ecosystem. However, it might seem that if Apple offered a technically better (for instance, more secure) payment system, it could be set up as the standard. Apple had signed with nearly a hundred banks, including Visa and Mastercard, in early 2015. One of Apple's advantages is that the system does not generate new intermediaries, in contrast to Google Wallet's virtual card, which is issued by Bancorp Bank. This means that someone has to pay an additional fee; Google for instance. With Apple's system, charges are made to the customers' own cards (Hibben 2015; Musil 2015).

Google's choice of mobile payment implies saving the information on these transactions in its storage servers. This gives the company valuable information

Table 5 GAFA. Examples of platforms

Market	Side 1	Side 2	
Smartphones	Consumers	Apps	Apple, Google
Payment mechanism	Buyers	Sellers	Apple, Google
Travel	Info about flights	Airline companies	Kayak (Google)
Apps	Developers	Users	AppStore, Android store
iTunes, Googlebooks	Consumers	Editors/studios	Apple, Google
Tablets	Consumers	Apps	Apple
Kindle	Readers	Books	Amazon
Mall	Shoppers	Providers	Amazon
Search	Searches	Information	Google
Search	Advertising	Advertisers	Google, Facebook
Facebook	Readers	News	Instant articles (Facebook)
Communication	Individuals Brands	Individuals	Facebook

Source Own elaboration

that could be used for its own benefit. On the contrary, Apple points out that it would not save any information. Nor would Apple have information on the device account number assigned to the iPhone (Hibben 2015).

GAFA tend to be autonomous concerning basic issues such as energy, building electric power plants and buildings for data storage. The degree of openness of business systems, understood as the degree of external dependence, could be considered, for example, in terms of innovation and/or manufacture. It would doubtlessly be very high, since Apple, Amazon and Facebook develop their own software and applications and acquire hardware, so the ecosystem is not completely closed. It is actually impossible to close because there will always be some activity that is impossible or difficult to integrate, such as applications. They all, particularly Apple and Google, have an application platform which is open to developers which propose applications to make them available to users. However, Apple's case is interesting as it depends on its rival Samsung to supply strategic components of its devices such as microprocessors.

Timothy D. Cook pointed out: "the key thing for us, Steve, is to stay focused on things that we can do best and we can perform at a really high-level of quality that our customers have come to expect. And so we currently feel comfortable in expanding the number of things we're working on" (SA Transcripts 2014). In order to make the best items, they are going to have to depend on others, which are sometimes their competitors.

Keeping an ecosystem closed has its price. In effect, some components do not have the chance to grow as a result of opening up. However, the brand image, and the ecosystem as a whole, would become blurred. If all of Google's applications could be used on iPhones, it would be easier to sell more iPhones. Nevertheless, as Apple would not control the applications, some of them could cause problems, and this could affect the system as a whole. The ecosystem would have the features of a commodity system, which means the quality of the whole would be determined by its lowest quality element. Keeping this ecosystem closed is complicated.

6.1 Apple Versus Samsung

Components provision is one of Apple's weakest points, and which must constantly be solved. Apple is the firm most widely known for the idea and example of an ecosystem (devices, operating system services, applications) although it cannot make everything, for instance processors, screens, batteries, etc. Samsung does not have such a clearly defined ecosystem although its manufacture of basic components gives it a certain edge over its most direct competitors.

It was Samsung Electronics Co. that manufactured the microprocessors for iPhones and iPads. In mid-2013, Apple entered into a partnership with a Taiwanese firm—Taiwan Semiconductor Manufacturing Co (TSMC)—to begin manufacture of chips in 2014 (Luk 2014). In 2013, the A7 processor was made entirely by Samsung. However, in 2014, it made only 30 % of A8 due to TSMC. Finally, in 2015 Samsung once again becomes the main manufacturer for capacity and costs reasons (Jones 2014).

Samsung will continue to be the main processors supplier in 2015, which puts Apple in a difficult situation since Samsung is its main competitor. Ritala et al. (2014) analyse this *coopetition* relationship applying it to Amazon, which supplies software while also advising its direct competitors.

Everything was running smoothly until they became such fierce rivals. Apples future is now in danger as Samsung provides its basic components—processors, screens for iPhones and iPads and memories. For this reason, Samsung has first-hand knowledge of the technical features of these components, information on new product launches and therefore marketing strategies. This knowledge could be used to improve and/or accelerate the pace of innovation of these components to later use them for Samsung's own benefit. Apple intends to reduce this dependence by looking for other suppliers. However, this will be difficult, at least for a few years. In 2008, Apple considered manufacturing its own processors but gave up the idea. What it has managed to achieve is Samsung's agreement to keep components and devices separate so that there is no intercommunication between these fields.

Apple is particularly concerned about the screens, which are its brand image and most visible component. The iPhone 4 was launched in 2010 with the retina display screen manufactured by Samsung. Apple also tried to find other screen suppliers such as Sharp and Toshiba. However, the new third generation iPad was released in 2012 once again with a Samsung screen in spite of Apple's attempts to have it manufactured by Sharp.

Samsung does not completely dominate the smartphone processors market. There are others such as the South Korean Hynix, which entered into agreements with Samsung in mid-2013 to end disputes over patents and develop better microprocessors (Lee, 2013).

Samsung is interested in selling components to Apple, as these supplies totalled around 10 billion dollars in 2012. This figure is approximately a third of Apple's total spending on processors in 2013 (Reisinger 2015).

The last information to take into account is that Samsung has had a 3 % stake in Sharp since March 2013, which may diminish Apple's negotiating capacity with Sharp. This means that components, their design and manufacture outside Apple is another epigenetic area that should be studied to see where Apple's ecosystem is going in the future.

It is also important to note that aspects that are normally external to companies such as electricity must be integrated in the ecosystem because GAFA's enormous energy consumption is such they must integrate it.

Google is one of the biggest electric energy consumers. In 2011, its energy consumption was equivalent to that of Austin, Texas. Such consumption cannot fail to be a great concern and for this reason, Google has invested in solar panel companies and other sources of more environmentally friendly electricity than the conventional ones. What Google wants to exploit above all is electricity-related big data. The purchase of Nest digital thermostats allows it to know: "We know how many people are home during those times, (…) We know which homes have cooling, we can go through the data and we can say we believe we're a five megawatt power generator or 10 megawatts in this county. That's what we can deliver—50–60 % of the energy that was consumed in that window has been shifted away" (Olson 2013).

7 Crowd Economies?

When studying GAFA, the question arises about the enormous amount of data they handle, leading us to think of new types of economies linked to big data that could appear. Companies can achieve different types of economies. In addition to experience and innovation, companies can obtain cost reductions (different types of economies). The classic types of economies that have been considered to the present time are related to the economies of scale (William et al. 1982); although these are sometimes thought to exist in audiovisual production (Hopewell 2015) and with scope economies (in acquisitions, for instance) (Chatterjee 1986).

Although they are sometimes confused, synergy is different from scope economies because it means or may mean higher revenues while scope economies arise from cost reductions. As Iversen (1997) points out: "synergy is concerned with more than the cost of production (...). Where economies of scope deal with the reduced costs of joint production (i.e. resource sharing) vis à vis separate production, synergies are also about increasing revenue and reducing the need for investment". Having put forth these considerations, we might examine what types of economies are characteristic of GAFA ecosystems. Since there is initially no industrial multiplier effect, it is not possible to cite the existence of economies of scale as a characteristic.

Scope economies may also exist, and even more synergies, insofar as they share resources, symbols, etc. However, the most characteristic are what we could call symbiotic economies, given that all GAFA business activities complement each other. In the case of Apple, there is a relationship that can be considered symbiotic between the same company's devices and content. There is complementarity between the Kindle and electronic acquisition of content in the case of Amazon. In any case, what underlies the use of devices and services supplied by GAFA are big data, which can later be used to improve existing products and services and launch new ones. Shouldn't daily massive use of services make us think of something we could call crowd economies? (Evans and Forth 2015) In effect, these data may be used by the companies to launch new business activities which apparently have nothing to do with the activities commonly attributed to them. Thus, insofar as Apple has data on millions of people, it can launch mobile payment services systems very easily as the customers are already registered. Just as they access an application with a password, they could pay via a mobile device without having to do anything except enter the same password they use for the Apple ecosystem (iTunes, iCloud, applications, etc.).

According to Statista, Apple had approximately 600 million user accounts linked to its services in the third quarter of 2013, Amazon 224 million active user accounts and PayPal 137 (Statista 2013). These figures enable us to understand that Apple is well positioned to supply intermediation services for mobile payment systems; most of the 600 million users in its database have credit cards associated to their accounts.

8 Conclusions

The interest in studying GAFA is based on their being synonymous with the Internet and undergoing constant innovation. New growth vectors are currently being started—wearable technology, Internet of things, automated automobiles, mobile pay technologies— which form part of the big data field, and are going to boost GAFA's growth, although the problems will also increase. GAFA are integrating mobile technology as a new opportunity in their strategic approach. The use of mobile devices raises new challenges for all of the companies. Internet searches via mobile devices are increasing although only totalling 29 % in 2015, with a slightly lower percentage for total purchases via mobile devices. They all want to be well positioned in the field of mobile devices, but they cannot all achieve this at the same time. Regardless of the technology, slightly over 50 % of online sales are initiated outside search engines. This situation has Google at war with Amazon because online shopping searches for products that are bought most often generate more sales at the purchase sites.

Growth of GAFA is not infinite as there are factors that limit it. Google is facing a *balkanisation* of the spaces where it can be used. Google's search robots cannot enter Facebook so all the photographs and texts found inside it are not accessible. This means that it can be used in fewer spaces. On the other hand, certain searches begin with Amazon as the entry point, for example. When searching for bibliography or information on a certain book or product, people may enter directly in Amazon, ignoring Google. Music-related searches can be done on Spotify, which means Google is no longer the universal search engine. The above points illustrate that searches can be done on Amazon or Facebook so they can also be considered search engines.

GAFA are now at the intersection of many issues that may affect their growth. Facebook and Google obtain most of their revenues from advertising. The tendency to look for new ways to obtain revenue other than through advertising is becoming apparent. Sale of items—Facebook's virtual reality goggles or Google's buy button are cases in point, aiming to turn searches into sales.

They can offer advertisers in-depth knowledge of user preferences, enabling them to implement marketing strategies. However, privacy issues immediately arise and strict regulations to protect users mean that it is more difficult to use or sell knowledge and data mining results.

Innovation is essential for all companies. And even more so for GAFA, since clients always expect the "WOW Effect". The stiffer the competition, the greater the pressure to innovate. As each GAFA member's number of competitors rises, so does the pressure to innovate. It is impossible to maintain a fast pace of innovation that always achieves a surprise effect. Study of how innovation is organised within each GAFA group should show the rest of companies the path to manage human resources and design innovation. However, this is no easy task because one of the key features is secrecy.

The innovation carried out in one activity unequivocally affects the others within the same firm or ICT ecosystem. Google Glasses and Car, as well as Apple's iWatch, will generate new applications, which may lead to new innovations. However, these innovations may mean that mobile telephony networks should increase their transport and management of voice and data traffic, etc. Not surprisingly, telecommunications companies sometimes file for remuneration for the large investments they have to make to satisfy the growing demands caused by this traffic and generated by downloads and use of applications for mobile devices.

Now that they have existed for several years, one question that remains to be analysed is their growth and development from the management perspective. How it is possible to manage the growth of companies that become so large and in just a few years have thousands of highly qualified staff.

GAFA have based their growth and development on technology originating inside the companies. Apple created its own software and designs its hardware. Amazon designs software and the logistics for its warehouses and for other firms, etc. The question is now: Given the extremely fast pace of innovation and new fields of growth mentioned earlier that are opening up, is it possible to generate all the necessary knowledge internally? That is to say, won't we see increased competition between GAFA and external firms? And how will this affect GAFA?

Their global nature and daily use raise key questions regarding economics (growth strategies, existence of double markets, pace of innovation) as well as other issues such as patents wars, privacy, taxes, abuse of power, etc. They are global in nature, but regulation is not. Therefore, the USA and Europe monitor their behaviour separately—issues concerning income tax, disloyal competition. We have referred to their global nature although their activity on the Asian market is low, except for Facebook, 30 % of whose users are found in the Asia-Pacific zone, and Apple, whose iPhone sales are growing in the area. Penetrating the Asian market is not easy but may present a possible area for growth. However, the opposite could occur, and the Asian counterparts of GAFA could start growing on western markets. No big changes are expected concerning market unification on the Internet in the short term, although this could cause large-scale structural changes.

References

Bensinger, G. (2013). Amazon is developing smartphone with 3-D screen. http://online.wsj.com/news/articles/SB10001424127887324744104578473081373377170. Accessed 29 July 2015.

Borrow, J. (2014). Giant tech brands: Which one has your loyalty? *Technology researcher*. http://conversation.which.co.uk/technology/apple-android-amazon-microsoft-ecosystem/. Accessed 29 July 2015.

Chatterjee, S. (1986). Types of synergy and economic value: The impact of acquisitions on merging and rival firms. *Strategic Management Journal, 7*(2), 119–139.

Chibber, K. (2014). American cultural imperialism has a new name: GAFA. *Quartz*. http://qz.com/303947/us-cultural-imperialism-has-a-new-name-gafa/. Accessed 29 July 2015.

Clarysse, B., Wright, M., Bruneel, J., & Mahajan, A. (2014). Creating value in ecosystems: Crossing the chasm between, knowledge and business ecosystems. *Research Policy, 43*, 1164–1176.

Constine, J. (2015). Skip googling with Facebook's new add: A link mobile status search engine. http://techcrunch.com/2015/05/09/share-without-leaving/?ncid=rss#.fcgmno:tNXv. Accessed 29 July 2015.

Dilger, D. E. (2014). Apple's voracious appetite for acquisitions outspent Google in 2013. *Apleinsider*. http://appleinsider.com/articles/14/03/03/apples-voracious-appetite-for-acquisitions-outspent-google-in-2013. Accessed 29 July 2015.

Ducourtieux, C. (2014). La France esquisse des pistes pour faire payer plus d'impôts aux géants du Web. *Le Monde*. http://www.lemonde.fr/economie/article/2012/12/20/la-france-esquisse-des-pistes-pour-faire-payer-plus-d-impots-aux-geants-du-web_1808875_3234.html?xtmc=gafa&xtcr=64. Accessed 29 July 2015.

Efrati, A. (2013). Google works on launching retail stores. http://online.wsj.com/news/articles/
SB10001424127887323764804578312530021763450. Accessed 29 July 2015.

Eisenmann, T. R., Parker, G., & Van Alstyne, M. W. (2006). Strategies for two sided markets.
Harvard Business Review, 84(10), 92.

Elmer-DeWitt, P. (2014). Nine bar charts: Apple versus Amazon versus Google. *Fortune*. http://
fortune.com/2014/01/06/nine-bar-charts-apple-versus-amazon-versus-google/. Accessed 29
July 2015.

Evans, P., & Forth, P. (2015). Borges' Map. Navigating a world of digital disruption. The Boston Consulting Group. http://digitaldisrupt.bcgperspectives.com/?utm_
source=201505BORGES&utm_medium=Email&utm_campaign=Ealert. Accessed 29 July 2015.

Fabernovel (2013). Gafanomics: New economy, new rules. http://www.fabernovel.com/.
Accessed 29 July 2015.

Fransman, M. (2014). *Models of innovation in global ICT firms: The emerging global innovation
ecosystems*. Seville: European Commission Joint Research Centre Institute for Prospective
Technological Studies.

Haucap, J., & Heimeshoff, U. (2013). Google, Facebook, Amazon, eBay: Is the internet driving competition or market monopolization? Discussion paper, 83, 7. Düsseldorf Institute for
Competition Economics (DICE).

Haucap, J., & Heimeshoff, U. (2014). Google, Facebook, Amazon, eBay: Is the internet driving competition or market monopolization? *International Economics and Economic Policy,
11*(1–2), 50.

Hibben, M. (2015). Apple pay vs. Google wallet: The rematch. *Seekingalpha*.
http://seekingalpha.com/article/2948836-apple-pay-vs-g. Accessed 29 July 2015.

Hopewell, J. (2015). Endemol Shine Iberia looks to economies of scale, synergies, digital, drama.
Variety. http://variety.com/2015/tv/global/endemol-shine-iberia-looks-to-economies-of-scale-
synergies-digital-drama-1201452285/. Accessed 29 July 2015.

Internet Live Stats. (2015). http://www.internetlivestats.com/one-second/#google-band. Accessed
29 July 2015.

Iversen, M. (1997). *Concepts of synergy: Towards a clarification*. Working paper for the DRUID-
seminar. Department of Industrial Economics and Strategy, Copenhagen Business School.

Jhonsa, E. (2015). Zuckerberg downplays Facebook/Google rivalry. Seeking Alpha.
http://seekingalpha.com/news/2395076-zuckerberg-downplays-facebook-google-rivalry.
Accessed 29 July 2015.

Jones, L. (2014). Samsung to make 80 % of Apple's processors in 2015. http://www.mobileburn.
com/23799/news/samsung-to-make-80-of-apples-processors-in-2015. Accessed 29 July 2015.

Juste, M. (2014). Facebook presenta HTC First, el primer móvil con la red social completamente
integrada. *Expansión*. http://www.expansion.com/2013/04/04/empresas/digitech/1365096005
.html. Accessed 29 July 2015.

Kelly, E. (2015). Introduction: Business ecosystems come of age part of the business. Deloitte
University Press. http://dupress.com/articles/business-ecosystems-come-of-age-business-
trends/. Accessed 29 July 2015.

Kokalitcheva, K. (2015). Whatsapp had 700 million users who sent 30,000 million messages per
day. *VentureBeat*. http://venturebeat.com/2015/01/06/whatsapp-now-has-700m-users-send-
ing-30b-messages-per-day. Accessed 29 July 2015.

Lawson, S. (2014). The company dropped its cluster design for a network fabric with smaller
server pods. *Computer world*. http://www.computerworld.com/article/2847864/facebooks-
iowa-data-center-goes-modular-to-keep-pace-with-growth.html. Accessed 29 July 2015.

Lee, M. J. (2013). Samsung wins over rival, secures chip supply. http://blogs.wsj.com/
digits/2013/07/03/samsung-wins-over-rival-secures-chip-supply/. Accessed 29 July 2015.

Lessin, J. E. (2013). Apple finds it difficult to divorce Samsung. http://online.wsj.com/article/SB1
0001424127887324682204578513882349940500.html. Accessed 29 July 2015.

Luk, L. (2014). TSMC starts shipping microprocessors to Apple. http://www.wsj.com/
articles/tsmc-starts-shipping-microprocessors-to-apple-1404991514. Accessed 29 July 2015.

Makinen, S. J., & Dedehayir, O. (2012). Business ecosystem evolution and strategic considerations: A literature review. Presented in 18th Engineering, Technology and Innovation ICE International Conference (p. 2).

Mannes, J. M. (2014). Apple iWatch: A big deal or not? *Seekingalpha*. http://seekingalpha.com/article/2120583-apple-iwatch-a-big-deal-or-not. Accessed 29 July 2015.

MarketWatch (2014). Amazon steals some of Netflix's thunder by scoring 24 for its prime instant video service. http://blogs.marketwatch.com/thetell/2014/04/01/amazon-steals-some-of-netflixs-thunder-by-scoring-24-for-its-prime-instant-video-service/. Accessed 29 July 2015.

Marous, J. (2014). Google, Apple, Facebook and Amazon should terrify banking. *The financial brand*. http://thefinancialbrand.com/41484/google-apple-facebook-amazon-banking-payments-big-data/. Accessed 29 July 2015.

Mars, M. M., Bronstein, J. L., & Lusch, R. F. (2012). The value of a metaphor: Organizations and ecosystems. *Organizational Dynamics, 41*, 271–280.

Martin, J. (2014). Après Facebook, Google rachète à son tour un fabricant de drones. *Lemondeblogs*. http://siliconvalley.blog.lemonde.fr/2014/04/15/apres-facebook-google-rachete-a-son-tour-un-fabricant-de-drones/. Accessed 29 July 2015.

Matthies, A., Stephenson, A., & Tasker, N. (2015). The concept of emergence in systems biology. www.stats.ox.ac.uk/__data/assets/pdf_file/0018/3906/Concept_of_Emergence.pdf. Accessed 29 July 2015.

Moore, J. F. (1996). *The death of competition: Leadership & strategy in the age of business ecosystems*. New York: Harper Business.

Moore, J. F. (1998). The rise of a new corporate form. *Washington Quarterly, 21*(1), 168.

Moore, J. F. (2015). Business ecosystems and the view from the firm. *The Antitrust Bulletin*. http://cyber.law.harvard.edu/blogs/gems/jim/MooreBusinessecosystemsandth.pdf. Accessed 29 July 2015.

Moritz, E. F. (2015). Modelling a business ecosystem in/for emerging industries. innovationsmanufaktur. http://www.mobilise-europe.mobi/modelling-a-business-ecosystem. Accessed 29 July 2015.

Musil, S. (2015). Amazon to fold its mobile-wallet app beta on wednesday. *Cnet*. http://www.cnet.com/news/amazon-to-fold-its-mobile-wallet-app-beta-on-wednesday/. Accessed 29 July 2015.

Olson, P. (2013). Nest gives Google its next big data play: Energy. *Forbes*. http://www.forbes.com/sites/parmyolson/2014/01/13/nest-gives-google-its-next-big-data-play-energy/. Accessed 29 July 2015.

Pereira, S. (2014). Facebook's oculus acquisition: Fundamentals, rewards and risks. *Seekingalpha*. http://seekingalpha.com/article/2112873-facebooks-oculus-acquisition-fundamentals-rewards-and-risks?isDirectRoadblock=false&uprof=11. Accessed 29 July 2015.

Perrot, A. (2011). Le numérique: Enjeux des questions de concurrence. *Concurrences, 3*, 1–6.

Pilinkienė, V., & Mačiulis, P. (2014). Comparison of different ecosystem analogies: The main economic determinants and levels of impact. Presented in 19th International Scientific Conference Economics and Management 2014 (ICEM-2014), Procedia Social and Behavioral Sciences (pp. 365–370), November 26, 2014.

Poulter, S. (2014). Engineer from Somerset farm is expert at centre of Facebook bid to get world's most remote regions online using drones and satellites. *Dailymail*. http://www.dailymail.co.uk/sciencetech/article-2592035/Facebook-use-solar-powered-British-drones-earths-remotest-spots-internet.html. Accessed 29 July 2015.

Rachman, G. (2015). The political storm over the Googleplex. *Financial Times*. http://www.ft.com/cms/s/0/b2eeb470-ecca-11e4-b82f-00144feab7de.html#axzz3nOOLBGb0. Accessed 29 July 2015.

Rebillard, F. (2010). Les intermédiaires de l'information en ligne. *Inaglobal*. http://www.inaglobal.fr/numerique/article/les-intermediaires-de-linformation-en-ligne/. Accessed 29 July 2015.

Reisinger, D. (2015). Apple, Samsung to remain bedfellows for next iPhone. *Cnet.* http://www.cnet.com/news/apple-samsung-still-bedfellows-as-they-eye-next-iphone-report-says/. Accessed 29 July 2015.

Ritala, P., Golnam, A., & Wegmann, A. (2014). Coopetition-based business models: The case of Amazon.com. *Industrial Marketing Management, 43*(2), 236–249.

Robinson, J. (2011). Google needs television industry will be message at Edinburgh. *The Guardian.* http://www.guardian.co.uk/media/2011/aug/21/google-needs-television-industry-edinburgh. Accessed 29 July 2015.

Rong, K., Guangyu, H., Jie, H., Rufei, M., & Yongjiang, S. (2013a). Business ecosystem extension: Facilitating the technology substitution. *International Journal of Technology Management, 63*(3–4), 268–294.

Rong, K., Guangyu, H., Jie, H., Rufei, M., & Yongjiang, S. (2013b). Linking business ecosystem lifecycle with platform strategy: A triple view of technology, application and organization. *International Journal of Technology Management, 62*(1), 75–94.

Rong, K., Guangyu, H., Yong, L., & Yongjiang, S. (2015). Understanding business ecosystem using a 6C framework in Internet-of-Things-based sectors. *International Journal of Production Economics, 159*, 41–55.

Rysman, M. (2009). The economics of two-sided markets. *Journal of Economic Perspectives, 23*(3), 125–143.

SA Transcripts. (2014). Apple's CEO discusses F2Q 2014 results: Earnings call transcript. *Seekingalpha.* http://seekingalpha.com/article/2159213-apples-ceo-discusses-f2q-2014-results-earnings-call-transcript. Accessed 29 July 2015.

Saviolo, P. (2014). Amazon: Ready for the smartphone and tablet market? *Seekingalpha.* http://seekingalpha.com/article/2058423-amazon-ready-for-the-smartphone-and-tablet-market. Accessed 29 July 2015.

Schmidt, E. (2011). Mactaggart lecture. *The Guardian.* http://www.theguardian.com/media/interactive/2011/aug/26/eric-schmidt-mactaggart-lecture-full-text. Accessed 29 July 2015.

Simon, P.H. (2011). *The age of the platform: How Amazon, Apple, Facebook, and Google have redefined business.* Motion Publishing.

Somaiya, R., Isaac, M., & Goel, V. (2015). Facebook may host news sites content. *The New York Times.* http://www.nytimes.com/2015/03/24/business/media/facebook-may-host-news-sites-content.html?_r=1. Accessed 29 July 2015.

Statista. (2013). User base of Apple, Amazon and Paypal as of 3rd quarter 2013, ranked by user accounts. http://www.statista.com/statistics/279689/apple-amazon-and-paypals-user-base/. Accessed 29 July 2015.

Statista. (2014a). Walmart's net sales worldwide from 2006 to 2015. http://www.statista.com/statistics/183399/walmarts-net-sales-worldwide-since-2006/. Accessed 29 July 2015.

Statista. (2014b). Leading social networks worldwide as of November 2015, ranked by number of active users. http://www.statista.com/statistics/272014/global-social-networks-ranked-by-number-of-users/. Accessed 29 July 2015.

Storage Servers. (2013). Facts and Stats of World's largest data centers. https://storageservers.wordpress.com/2013/07/17/facts-and-stats-of-worlds-largest-data-centers/. Accesed 29 July 2015.

Subramanian, C. (2014). Google's plans for a store may be more like a museum. http://tech.fortune.cnn.com/2014/03/13/googles-plans-for-a-store-may-be-more-like-a-museum/. Accessed 29 July 2015.

Travlos, D. (2013). Importance of being a platform: Apple, LinkedIn, Amazon, eBay, Google, Facebook. *Forbes.* http://www.forbes.com/sites/darcytravlos/2013/02/26/importance-of-being-a-platform-apple-linkedin-amazon-ebay-google-facebook/. Accessed 29 July 2015.

William, J., Baumol, J. C., Panzar, R. D., Willig, E. E., Bailey, D. F., & Fischer, D. (1982). *Contestable markets and the theory of industry structure.* New York: Harcourt Brace Jovanovich Inc.

World Economic Forum. (2015). Digital ecosystem convergence between it, telecoms, media and entertainment: Scenarios to 2015 http://www.weforum.org/reports/digital-ecosystem-convergence-between-it-telecoms-media-and-entertainment-scenarios-2015. Accessed 29 July 2015.

The Digital Ecosystem: An "Inherit" Disruption for Developers?

Jorge Vega, Jon Mikel Zabala-Iturriagagoitia
and José Antonio Camúñez Ruiz

1 Preface

First of all, however knowledgeable about programming or technology in general you might be, we would like all our readers to feel comfortable with the following text. I'm not a PhD or university lecturer but Jorge Vega, a developer, so I'm letting you know from the word go that this is not going to be like any of the other chapters you might have read in this book. We'll only be including a few bibliographical references and most of them will be links. We'll do our best to use language similar to that in the rest of the book even though academic terms and developers' jargon have little in common. We'd like this chapter to be easy reading, written in a straightforward style, where we can speak directly to the reader, like we would at a talk or debate or just a friendly chat in a coffee shop.

As you may have noted this different style is already present in the title of the chapter. We have included the word "inherit" in it, in a nod to the language of developers and to the meaning of the word in relation to the Darwinian jargon of inherited characters. The aim of this chapter is to show developers' vision of the dynamics that occur in the Internet ecosystem. Up to this point, you've read about

J. Vega (✉)
Senior Front-End Developer, Bilbao, Spain
e-mail: jvega300@gmail.com

J.M. Zabala-Iturriagagoitia
Deusto Business School, University of Deusto, Donostia-San Sebastian, Spain
e-mail: jmzabala@deusto.es

J.A.C. Ruiz
Department of Applied Economics I, University of Seville, Seville, Spain
e-mail: camunez@us.es

© Springer International Publishing Switzerland 2016
M. Gómez-Uranga et al. (eds.), *Dynamics of Big Internet Industry Groups and Future Trends*, DOI 10.1007/978-3-319-31147-0_5

149

the characteristics of the big business groups that dominate said ecosystem. But in this chapter, we'd like to take a look at the other side of the coin and analyse what is happening in the world of developers, what they do and how the Internet is seen from the thousands of start-ups that form it. Do developers think that there is a disruption that is the same or equivalent to what the big Internet companies create in other scopes? That's the question we will try to answer. We're fully aware that hundreds of thousands of developers and start-ups form this new ecology which is constantly expanding, so this should all be taken as the view of just one developer and may well not agree with that of many other digital developers or entrepreneurs.

Like all good computer freaks (and proud of it!) we developers who are now over 35 have been tremendously influenced by the great movie sagas and science fiction movies. And like all good trilogies, the chapter will start with a story, which we'll gradually set in its proper context. The second part of the chapter includes Sects. 2 and 3. Section 2 pays special attention to the evolution of programming languages over the last decades because it is key to understanding the evolution of developers' logic. It also gives an overview of the different developer profiles. Section 3 covers developers' growing empowerment in the Internet ecosystem and the increasingly important role they play in it. Section 4 shows us a particular vision of the possible future scenarios that we might expect, from the perspective of developers and start-ups and their role in the Internet ecosystem as a whole. Section 5 centres on the opportunities that might arise from the emerging Big Data context and the role that developers may play in it. The chapter ends by analysing the weaknesses of the epigenetic (i.e. EED) approximation for studying developer-related dynamics in the previous scenarios and raises the possibility of their being studied from a quantum approach.

2 What Being a Developer Means

I went to a talk by Carlos Barrabés in San Sebastian about 15 years ago, during which he explained his experience with his online shop.[1] He talked about how he had hired some consultants and when he told them about his online business idea, they started picking it apart, and tried to change the direction of the project he had in mind. Finally, he got so fed up with snags and not being able to put his idea into practice that he decided to get rid of them and hire a developer, with whom he started his online business. And that very business has, with time, become a global success.

---

[1]Carlos Barrabés is an entrepreneur who was born in Benasque (Spain), a town in the middle of the Pyrenees in 1970. Barrabés ran a small mountain gear shop and set up the first online shop around 1994 (i.e. sale via the Internet) in Spain and one of the first in the world. See: http://www.barrabes.biz/.

As I listened to him, the question I asked myself, and would now like to ask you, is: who was the developer in that story? The first consultants, the developer (i.e. computer engineer) who set up the project that Barrabés had in mind, or Carlos himself? Besides laughing at the insults the speaker launched at computer engineers, the only thing I remember about that talk was coming out with the feeling that everybody is "a bit of" a developer. More than anything else, I remember how the speaker highlighted that everything was possible in the digital world and the most important thing is ultimately for each person to be engaged enough with the organisation they belong to so that they give their best to achieve an overall final result that satisfies all the parties concerned. Somewhat like what first year Economics or Business Administration students are taught in the lectures on the optimising individual in microeconomics.

To start with, I would like to get a few things straight and clarify what a developer is and isn't by pointing out the differences between developers and programmers. Programmers are the ones who write codes. Full stop. They don't give a care about the purpose of their code except for its function declaration. However, developer is a broader term. A developer is anyone who has a professional profile that requires defining and creating (i.e. modelling) a product or service, and doesn't have to be a programmer to do so. I realise that this idea is completely subjective and not everyone may agree, but.... I guess that's why people discuss these things, isn't it? That is the great advantage of science; you can have informed debate where as many points of view as there are disciplines converge (or collide). In any case, that term referring to disciplines is being questioned due to the prevailing interdisciplinary nature of today's world and the Internet ecosystem is no exception to this.

Here I am in 2015 writing this on a laptop that has an operating system (OS) with many programmes and routines where thousands of lines of code are processed in a second. Actually, this chapter is being written gradually using different platforms and devices that are synchronised. So depending on where I am, or how inspired I feel, and many other factors that I'm not going to mention here, there are days I write from my laptop and other days from my tablet and, when on the underground, from my smartphone. Sometimes, I even write on a paper napkin in a café and later include that digitally in the text or talking directly to any of the previous devices, depending on how illegible or legible the napkin might be when checking it at home. For instance, I used two editing software for this text, an online one and a desktop one. Both allow me to edit the same text without losing any information from various Android devices and my Mac, all 100 % compatible. As I mentioned, all these programmes or apps run on different hardware but thanks to online services, I can create the illusion that all my data magically appear everywhere. What's more, if the apps can run in a browser, we see that the information is available from almost anywhere (e.g. a terminal, a browser, a mobile OS). I never knew writing could be so difficult. When I was first asked to contribute to this book, I thought writing a chapter couldn't be too different from writing lines of code. One of the conclusions I've reached is writing code and writing a book or scientific articles are just about as different as developers and programmers.

Needless to say, the Internet is just one of the current areas of technological development where developers play the role they always did. Today's gurus might have forgotten that it took a group of developers to make an alarm clock just 25 years ago. Developers, however, don't just focus on developing code because hardware is essential. As Quintero (2015) pointed out, investment in hardware has grown exponentially since 2010. I imagine that with the connection leap from our machines to our communication terminal (i.e. the Internet of Things), this investment will get even bigger because a lot of the equipment we now have in the home will have to be updated.

For instance, in business information management, we are shifting from Enterprise Resource Planning (ERP)-based environments, which would be somewhat like business resources planning systems (i.e. automate the company's processes in any of its departments via different modules) to cloud Software as a Service (SaaS) systems, where the software is online and we connect to its management capabilities without having to use ultra-expensive machines to do so. That's how new business models arise in code generation or for companies that develop these technologies that didn't exist before such as[2]:

- **Infrastructure as a Service (IaaS)**: In this case, processing capacity (CPU) and storage are contracted. Our own applications can be deployed in this environment if we choose not to install them in our company to avoid costs or due to lack of knowledge. Servers manage them and all expenses become variable costs for customers, so that they only pay for what they use. Examples of IaaS could be Amazon's Elastic Compute Cloud (EC2) Microsoft's Azure.
- **Platform as a Service (PaaS)**: Here, an applications server (where our applications will run) and a database are provided so that we can install the applications and run them. You usually have to observe a series of restrictions to develop apps for a server, concerning programming languages for instance. An example of PaaS could be the Google App Engine.
- **Software as a Service (SaaS)**: This is what is commonly called "the cloud". It's an application for end users who pay a rent for the use of the software stored in said application. This means that users don't have to buy software, install it, set it up and maintain it, since all this is done these by the SaaS. Some examples of SaaS could be Google Docs, Zoho or Office365.

Yes! Whoever you are, reading this chapter, no matter what they have told you… there is not really a cloud! That image of something they say is called "the cloud" is nothing more than a room full of machines. This is nothing new. It's an old concept.

I often meet people (many of them are university lecturers) who define a group of services as something tangible (I mean "the cloud" concept). That really bothers me. Honestly, the people who should be educating Internet users and people in

---

[2]See: http://www.xatakaon.com/almacenamiento-en-la-nube/cuando-hablamos-de-la-nube-que-es-iaas-paas-saas (last access 11th October 2015).

general are pushing an empty misleading concept. There is one concept that has been essential my whole life, and that is learning. This is true at least for me and most of my colleagues. As I was searching for a little bibliography for this paragraph, I ran into the "cerebral plasticity" concept, which is nerve cells' ability to anatomically and functionally regenerate as a result of environmental stimuli. Learning is all about achieving the best functional adaptation to the environment. That's what epigenetics is all about, isn't it?

Let's take a short pause here. The learning process must necessarily be connected to teaching, for us to understand it easily. When an apps critic or a technology journal talks about the cloud, or online services, when a salesperson is going to sell services to a company, both parties (teacher and student, salesperson and customer) should do our best to understand exactly what we are selling and hiring. And that's what I mean when I said you can't sell empty concepts like "the cloud". They have to be correctly explained and neither party should be misled, especially in business relationships where resources are invested and expenses and profits are everyday matters.

So, why do we give things such ridiculous names instead of boosting and conveying existing knowledge? It's as if we said that five-year-olds are naughty, annoying, etc. when we know that it's simply not true. Nobody can stick a child in a category just because of their being a child. If we extrapolate this idea to users, the same goes for services. Not all of them can be put in the "cloud" category. There are different services, adapted to meet certain needs, although generic services do exist. That kind of scalability is what online services provide. And I say "online services" because they may vary widely. I'm not using the term "cloud" as if it were just one service.

So far, I have tried to explain how, in general, we have shifted from closed environments to being exposed to a vast information network to help readers understand the delocalisation of typical hardware in big companies (e.g. Microsoft Windows) has created new business models for the entire world. And in turn, we have somehow stopped wanting to understand how things work and what they are for. It is true that our field is moving faster than Fernando Alonso's McLaren (2015) but, as digital citizens, we should at least try to explain the advances in the most accurate, friendly and interesting way possible. And we should also attempt to understand what their purpose is or what they can really be used for, both from a personal as well as a business perspective. But as we said at the start of the chapter, our aim is to show how developers view the epigenetics carried out by the big business groups in the Internet ecosystem. From the time I was asked to take part in this book, epigenetic dynamics have been on my mind. Despite seeing indicators that lead me to believe that the world of developers will be affected by the epigenetic model on the one hand, the lack of truly disruptive elements makes me wonder if the analytical framework provided by EED is really the right way to analyse the reality of developers. On the contrary, an alternative model might be needed to understand the other side of the coin in the digital ecosystem.

This inner struggle started out with the personal experience I have had with Marketing and my vision of it. I don't think an actual disruption exists at the development level, but rather that the reality of developers follows evolutionary logic more similar to Darwin's (see Chapter "Introducing an Epigenetic Approach for the Study of Internet Industry Groups" in this book by Gómez Uranga, et al.). The reality of most developers is pretty far from the aggressive marketing used for products and services and counterposes the benefits of development itself, which is the Internet companies' progress and advance. I think where there is fierce disruption is at the economic, social, institutional and business level but not so much at the development level. So, these companies, and therefore their marketing, are nurtured and grow thanks to the hours devoted to development by many people who are still working with systems that are more conventional than the final product. This sounds like the description of a big company although it isn't so different from a start-up, except that its investments are infinitely lower in every sense except for development and that is where good developer performance can "change the world".[3] So, I do agree that the Internet ecosystem is a two-sided coin. On one side, we have the big multinationals in the ecosystem, which have been referred to as GAFA in this book (see Chapter "Epigenetic Economics Dynamics in the Internet Ecosystem" by Zabala-Iturriagagoitia et al.). And for that side of the coin to exist, the other side where the thousands of start-ups and millions of developers are found needs to be firmly established.[4] However, I'm not so sure that the analytical model we can use to explain the dynamics of some (i.e. GAFA), is the right one to explain the evolutionary dynamics of the others (i.e. developers).

3 Developers: Classification and Evolution

I didn't study biology. Or economics or engineering. Therefore, I have ideas about what I have read in the authors' conversations that may not agree with theirs. And I accept that anybody should feel free to criticise and disagree with the views in this chapter as well. As Kuhn (1962) pointed out, scientific paradigms are often challenged by new models, leading to scientific breakthroughs. That's why science is distinctive for fostering widespread debate.

I understand an epigenetic dynamic to be evolution that occurs in response to the environment, particularly when it behaves like a high velocity environment where sudden disruptive changes take place. We are what we are thanks to our grandparents' diet, the pollution where they lived, the jobs they had and the same

---

[3]All you have to do is see this clip from "Silicon Valley", a series I highly recommend. View it at: https://www.youtube.com/watch?v=J-GVd_HLlps.

[4]According to IDC's 2014 Worldwide Software Developer and ICT-Skilled Worker Estimates (IDC 2013), the total number of software developers in the world is about 18.5 million. Around 11 million of those would be professional developers, and 7.5 million would be hobbyists (i.e. coders building software in their spare time for their personal entertainment, student developers, contributors to free and open-source software projects, and unfunded entrepreneurs).

goes for our parents; genetic legacy after genetic legacy shared with other humans and the environment. Bearing this concept in mind and getting back to our subject, the first thing we should do is take a look at the history of computer code. Understand what it was for, the environment where it worked, its aims and above all, what developers were doing at the time. In other words, we need to know the developers' original DNA.[5]

Looking back, we find that (programming) languages were originally used for a series of very specific activities with a minimum diffusion capacity in a very highly controlled ecosystem. As time passed, technological progress has made the ecosystem bigger, having reached today's global Internet. Several years ago, every developer had to learn many languages, which ranged from the simplest to the most complex. I think we all began programming in C. Or at least we had to go through that stage to learn the basics. A terminal.... and then we went right on to print code on a screen! Simple. From C, we moved on to C++, Turbo C, etc. They were all evolutions based on the same (relatively simple) pattern. Watch out, though. If you are a developer and didn't start with the above, I think it's a little dangerous because you lack the basics, the essence of the simple. There is nothing like seeing your first code printed on a terminal.

Taking into account how these languages evolved (Table 1), we can see that some "ghettoes" were created where languages were being embedded. Either because of the need to adapt the hardware to the language or because the language had to evolve thanks to the progress made in the hardware.

From the above list, we see how the first languages centred on computation in closed environments with very limited objectives. We should bear in mind that computational power was established thanks to the development of smaller transistors and chips, so we can put more in the same physical space. Moore's Law has been in force for no less than 50 years. Imagine an upward curve from 0 to 100 where we easily understand that speed has undergone an exponential increase. Going back to languages, the newest ones are oriented, and later adapted, to work on open systems (i.e. the Internet) and more specifically, on clients and servers.

We are witnessing a key moment in history, although we often don't see it that way. Mobility is changing everything. However, it requires the device concerned to have a good battery that won't run down because of the processor, without overlooking continuous connection to the Internet, which is causing some changes on servers' infrastructures (see Chapter "4G Technology: The Role of Telecom Carriers" in this book by Araujo and Urizar). This calls for a language that is light and can be extended to any hardware support.

I'm going to explain what the client/server parts are, for those readers who are not familiar with this. It's important to understand this point to be able to continue from this point and grasp what you are actually doing when you are holding

---

[5]This stage would be equivalent to what Gómez Uranga et al. identify in Chapter "Introducing an Epigenetic Approach for the Study of Internet Industry Groups" in this book, where they introduce the three-stage methodological approach of the EED, as the "Analysis of the environment and identification of the genomic instructions which are transmitted over time".

Table 1 History and influence of programming languages

Language	Named after	Year	Creator	General purpose	Primary uses	Used by
Fortran	The IBM Mathematical Formula Translating System	1957	John Backus (IBM)	High-level. For numeric and scientific computing (as an alternative to assembly language). Oldest programming language (still used today)	Supercomputing applications (e.g. weather and climate modelling, animal and plant breeding, computational science functions)	NASA
Lisp	List processor	1958	John McCarthy (MIT)	High-level. For mathematical notation. Several new computer science topics (e.g. tree data structures, automatic storage management, dynamic typing, and self-hosting compilers)	Algorithmic language development, air defense systems	Etsy uses Clojure, a dialect of Lisp[a]
Cobol	Common Business-Oriented Language	1959	Short Range Committee (SRC)	High-level. Primarily for business computing. First programming language to be mandated by the US Department of Defense	Business software (e.g. finance and administration systems), but also banks, insurance agencies, governments, and military agencies	Credit cards, ATMs
BASIC	Beginner's All-purpose Symbolic Instruction Code	1964	John George Kenny and Thomas Eugene Kurtz of Dartmouth (SRC)	High-level. Designed for simplicity. Popularity exploded in the mid-1970s with home computers. Early computer games were often written in Basic, including Mike Mayfield's Star Trek	Home computers, simple games, programmes, utilities	Microsoft's Altair BASIC, Apple II

(continued)

Table 1 (continued)

Language	Named after	Year	Creator	General purpose	Primary uses	Used by
Pascal	After French mathematician/physicist Blaise Pascal	1970	Niklaus Wirth	High-level. For teaching structured programming and data structuring. Commercial versions widely used throughout the 1980s	Teaching programming Object Pascal, a derivative, is also commonly used for Windows application development	Apple Lisa (1983), Skype
C	Based on an earlier language called "B"	1972	Dennis Ritchie of Bell Labs	Low-level. Created for Unix systems. Currently the world's most popular programming language. Many leading languages are derivatives, including C#, Java, JavaScript, Perl, PHP, and Python	Cross-platform programming, system programming, Unix programming, computer game development	Unix
Ada	After Ada Lovelace, inventor of the first programming language	1980	Jean Ichbiah	High-level. Derived from Pascal. Contracted by the US Department of Defense in 1977 for developing large software systems	US Department of Defense, banking, manufacturing, transportation, commercial aviation	NSTAR, Reuters, NASA, subways worldwide
C++	Formerly "C with Classes"; ++ is the increment operator in C	1983	Bjarne Stroustrup	Intermediate-level, object-oriented. An extension of C, with enhancements such as classes, virtual functions, and templates	Commercial application development, embedded software, server/client applications, video games	Adobe, Google Chrome, Mozilla Firefox, Microsoft Internet Explorer

(continued)

Table 1 (continued)

Language	Named after	Year	Creator	General purpose	Primary uses	Used by
Objective—C	Object-oriented extension of C	1983	Brad Cox and Tom Love of Stepstone	High-level. Expanded on C, adding message-passing functionality based on Smalltalk language	Apple programming	Apple's OS X and iOS operating systems
Perl	–	1987	Larry Wall of Unisys	High-level. Created for report processing on Unix systems. Today it's known for high power and versatility	Computer-generated imagery, database applications, system administration, network programming, graphics programming	IMDb, Amazon, Priceline, Ticketmaster
Python	For British comedy troupe Monty Python	1991	Guido Van Rossum of CWI	High-level. Created to support a variety of programming styles and be fun to use	Web application, software development, information security	Google, Yahoo, Spotify
Ruby	The birthstone of one of the creator's collaborator	1993	Yukihiro Matsumoto	High-level. A teaching language influence by Perl, Ada, Lisp, Smalltalk, etc. Designed for productive and enjoyable programming	Web application development, Ruby on Rails[b]	Twitter, Hulu, Groupon
Java	For the amount of coffee consumed while developing the language	1995	James Gosling of Microsystems	High-level. Made for an interactive TV project. Cross-platform functionality. Second most popular language (behind C)	Network programming, web application development, software development, Graphical User Interface development	Android OS/apps

(continued)

Table 1 (continued)

Language	Named after	Year	Creator	General purpose	Primary uses	Used by
PHP	Personal Home Page	1995	Rasmus Lerdorf	Open-source. For building dynamic web pages. Most widely used open-source software by enterprises	Building/maintaining dynamic web pages, server-side development	Facebook, Wikipedia, Digg, WordPress, Joomla
JavaScript	Final choice after "Mocha" and "LiveScript"	1995	Brendan Eich of Netscape	High-level. Created to extend web page functionality. Dynamic web pages use for form submission/validation, interactivity, animations, user activity tracking, etc.	Dynamic web development, PDF documents, web browsers, desktop widgets	Gmail, Adobe Photoshop, Mozilla Firefox

Source Veracode (2013)

[a]Etsy is an online marketplace for artists, designers and crafters, so that peer-to-peer e-commerce relationships focused on handmade or vintage items and supplies, as well as unique factory-manufactured items can be exchanged. Clojure is a general-purpose programming language with an emphasis on functional programming. It runs on the Java Virtual Machine, Common Language Runtime and JavaScript engines

[b]Ruby on Rails is a web application framework running on the Ruby programming language under MIT license

your terminal and you run actions on the applications. Besides, if something can be explained and someone learns, as we said earlier, so much the better, so here I will put those ideas into practice.

Readers who already know this can skip this paragraph. Imagine how often we open a browser and key in our favourite newspaper. Services represent the client, or even better, the programming of n elements, which enables the browser to use it as the language to communicate with whatever there is on the Internet. All the programming runs on my machine and the device uses software in the language that best suits to it so as to process the information got from the Internet. The server part covers processes that run on several machines across the world which work to supply us the information that we, as clients, ask for. We've all seen how the computer slows down when we load a webpage. That is because the information being interpreted as it reaches us. It's actually being "translated, understood and displayed in a comprehensible way", thanks to the code that has to run. When the computer slows down, that means there are problems or errors, and depending on the operating system, this will determine how angry we will get.

Imagine Mariano Rajoy, Pedro Sánchez, Albert Rivera and Pablo Iglesias in a meeting, or any other politician from your country. Since they are incapable of understanding each other, they have a translator on hand. They all speak Spanish but don't understand each other so suppose a question like "How is Spain doing?" comes up, which would be the information needed from the server part, the interpreter. Depending on its programming, it will make these deliver a response, according to how the information has been interpreted. I'll leave their responses to your imagination.

These two concepts, client and server, are necessary to understand the next section. As developers, please allow me to focus on the developers' ecosystem from my perspective as front developer. New specific jobs have been created which depend on the working model we embed in our online application (I say application so as not to use webpage or online business) or the language to execute on the client/server part.

If we take a look, there are currently many development frameworks (we understand frameworks as utilities) to develop a web application, for instance, what we know as a single-page application (SPA). Backbone, AngularJS, Ember or React, are all based on Javascript, a language from 1995 (see Table 1), but these evolutions enable a native browser format for their interpretation. However, the most important point is their ability to work on the "Model view controller" where:

- **The Model**: shows the information with which the system operates. It manages all access to such information, both queries and updates, also implementing access privileges that have been described in the application specifications (i.e. business logic). It sends to the "view" the part of the information that is requested for viewing at each moment (usually by a user). Information access or manipulation requests reach the "model" via the "controller".

- **The Controller**: responds to events (usually user actions) and sends commands to the "model" when some request is made concerning the information (e.g. edit a document or an entry on a database). It can also send commands to its associated "view" if a change in the way the model is displayed is requested (e.g. displacement or scroll for a document or the different entries in the database). So we could say the "controller" acts as an intermediary between the "view" and the "model" (see Middleware).
- **The View**: generates a presentation based on changes in the "model" (information and business logic) in a format that can interact (normally the user interface). It therefore asks said "model" for the information it should show as output.

When we load information on the browser, we wait to obtain a result. We're not aware of this action and are outside the application structure that runs internally both on the client/server. Increasingly, lighter workloads for the client and greater access to the processes that run on the server are being searched for. Since the latter are online, they may have higher level of computation. Although, as I said, for us (as users) it is invisible. We are only aware of our webpage's load time.

A further step in recent years, the arrival of Node.js,[6] is changing the very architecture of the applications, which is always done under the model view controller. Node.js enables you to have Javascript in the server part and not only in the client when this was and has been exclusively used in the client part. Disruptive? It may be, but it's applying a language in a familiar environment, which changes many things but is perhaps just one more step rather than a huge leap. The change is taking place gradually, the runtime environments have not been made obsolete by the new ones.

Having seen the intangible part of the code, I'd like to make a few remarks about programmers (i.e. who are not developers). In order to offer a possible classification of the many programmer profiles, I'd like to reflect on the how they are characterised (besides the fact of being fantastic technicians). This classification and the characterisation that goes with it may sound ironic, but if you have worked with code I'm sure you can identify the following descriptions:

- **The "Benito" programmer (From the TV series Benito y Compañia)[7]**: Maybe 80 % (Here's to Pareto!!!) of the programmers for SMEs and big companies fit in this category. They focus on debugging and solving problems that require speedy solutions, without needing a precise code. These programmers are vital to many semi-public companies and you can usually tell who they are when they ask the question: "well, it works, doesn't it?".

---

[6]Node.js is an open-source, cross-platform runtime environment for developing server-side web applications.

[7]Also known as the Benito Lopera Perrote (for Spanish readers) or the Mac Gyver (for international readers) of programming.
See: http://www.imdb.com/character/ch0169507/?ref_=nm_flmg_act_11 (last access 12th October 2015).
See: http://www.imdb.com/title/tt0088559/ (last access 12th October 2015).

- **The "Perfectionist" programmer**: These have a twin type in the design world. They can devote hours and hours, which ups the project budget. And when they deliver the project, you'd better not touch anything. They need to be kept close by on a short leash. Quoting the French writer Alphonse Karr (1808–1890), we could say this type of programmer "makes everything around him perfect, but does not strive to perfect himself". Also known as the Sheldon Cooper of programming.[8]
- **"By and by" programmer**: These are good programmers but are worn out from years of work and need to read a couple of sports dailies before they get down to doing anything job-related. They always say they'll have everything back to you "by and by". This doesn't mean they're not good at development, just that they don't feel like doing it when you need it. And then they are capable of creating all you need in record time. Also known as the Usain Bolts.[9]
- **"Technophile" programmer**: They always have the latest thing, on their cell phones, watches, etc., and know thousands of theory concepts in depth but are incapable of developing anything on their own. They are often seen as flies that flit around other technicians, distracting them and proposing ideas without knowing exactly what is being developed. Also known as the Antonio Recio of the wholesalers (for our Spanish readers)[10] or Milhouse Van Houten (the Simpsons).[11]

Of course, there are many other profiles, and hybrids of the previous ones, but I'm sure you've run into some of them if you've been working with programmers for a while. To tell you the truth, technicians (i.e. programmers) have also learned to evolve. You no longer have to be a Linux freak to be a good programmer. Actually, I really think that computer freaks are evolving towards a greater social awareness. Maybe because they understand the capacity of what they are developing or what they could actually achieve with their tools. In recent years, "hacktivists" are playing a bigger role and are more important thanks to the groups formed across the world.

I'd like to give some visibility to hacktivism (the term comes from the combination of hacking and activism) in just a couple of paragraphs. I won't go into any in-depth explanations since that is not the aim of this chapter. This socio-digital awareness is an indicator of what used to be the source of continuous jokes about their having no sex life but is now becoming significant. These movements are based on the capacity to join together digitally, and plan digital or social actions. Whether you agree with their activity or not is up to you, I only intend to show how developer groups have evolved. Epigenetics? I don't think we can deny that political and social concerns drive these groups to organise and carry out activities. But maybe what is most important is that, thanks to the developers in these

---

[8]See: http://www.imdb.com/character/ch0064640/ (last access 12th October 2015).

[9]See: http://usainbolt.com/bio/ (last access 12th October 2015).

[10]See: http://thecommentsection.org/viewarticle.php?id=5025 (last access 12th October 2015).

[11]See: http://www.imdb.com/character/ch0003035/ (last access 12th October 2015).

groups, there are tools that can make anybody a hacktivist one way or another, simply by being aware of the movement's ideas. Disruption? In my opinion, this is rather a gradual process that evolves hand in hand with society.

The broad field of development is generating an increasing number of new physical development nuclei causing the office ecosystem to begin to lose its original role. And that is where the business leaders-developers (i.e. entrepreneurs) that embarked on a path years ago came from, some of whom have been successful. However, the majority have had to accept their essential failure of not having become millionaires.

So, let's get back to languages after those brief comments on basic issues. Technological evolution has created several changes of direction. Nowadays, (practically) any programmer can create something with a beginning and an end, according to his needs, preferences or life projects. However, another technician may appear on the scene parallel to this and take advantage of part of the knowledge (i.e. code) created by the first person and bring out something entirely new or even focus that same code on different services. The number of opportunities depends on the enormous amount of available code. I often think the digital world is like the world of fashion, where cloth, designs, cuts and materials are all there, but each designer is capable of creating something totally different and sometimes unique. I believe that when language no longer depends solely on hardware, but on the service it is meant to provide, things will change in the development world.

4 Developer Empowerment

If you're a developer, of any type, and you have the mindframe to set up on your own with your technical knowledge, I feel sorry for you. There was a time for that. But today it's a lot more complicated. Not long ago, when they called you to do an online project, you just had to create a digital image like a showcase of what the company supplied. If you want to do something decent today, you have to upload the company webpage on the Internet, which means not only putting in their image but all of their organisation and customer management services.

If you want to do that on your own, I think your physical location is vital. The first thing to take into account is that living and working in a village or small town is not the same as in a big city. It's important to understand that the location of technology poles or clusters is important even if the network is global. Obviously, this also depends on your business aspirations. Several factors come into play here. For example, the first two authors of this chapter come from a town called Ermua between Bilbao and San Sebastian, in the Basque Country. Our town's industrial fabric is based on automotive parts production. And like ours, other towns nearby have a similar industrial fabric. European structural funds prompted many of these municipalities to develop strategies for conversion, we could say, toward manufacturing technologies (i.e. generally known as Industry 2.0 which has currently become the so-called Industry 4.0). Thanks to our "great" politicians (another key factor), a lot of projects with the same focus suddenly appeared in the political arena. They were incapable of coordinating these projects or of thinking about the overall development and well-being of the area. The result was semi-occupied buildings and struggles to attract business to the area. I am making this remark because I'm sure there are excellent developers, who, due to the circumstances, haven't had the chance to move forward with all their potential and have simply been swallowed up by the ideas of political leaders. And I repeat, instead of having a global vision, finding out what is involved in becoming a technology pole (there are scientists and researchers such as the authors of this book who have spent decades working on this subject) and finding local initiatives to drive development, these politicians wanted to create a second-rate Silicon Valley. As if it were something as easy as Grandma's paella recipe.

I said earlier that some years ago, it was not difficult to work as what we normally understand as a "web developer" if you had some knowledge and a certain amount of self-confidence. The term was certainly unfortunate, it was really like stuffing all of a developer's abilities in a box and giving it a kick to mix everything together. Most of the demand for web developers was pretty superficial, and didn't look to go much further than virtual showcases of the contracting company's true business activity. However, in most cases, they lacked business logic, without any possibility of getting any real productivity from being online in spite of developers who often wanted to suggest initiatives that the companies did not (or chose not to) understand. A few clever individuals managed to become programmer-designers and carried out projects that we find very old-fashioned today.

In today's context, when developers aim to become entrepreneurs, they run into "limited" entry barriers to the ecosystem (aware that legislation plays a key role in this sense and varies according to the country). However, the main obstacle they face when trying to grow in the Internet ecosystem is the fight to survive (i.e. what Moore 1993, referred to as "predators and prey").

For those who are still restless and haven't wanted to become obsolete due to the new technologies coming from big Internet companies, web design has become a specialist field. For those who haven't, however, the job market is continuing to shrink. So, if you still don't know several languages and don't make a profit (or don't know how to), on what you create, you need to get out of the chair and get your brain going. It will be harder and harder to get customers, unless you are extremely lucky (i.e. meaning they have no idea of the Internet's possibilities and make do with what you offer, which is nothing because there are tools that can do your job).

There used to be a saying: look for something in the real world and create it in the virtual world cheaper and more profitably. But that is no longer the case. Everything changed when some developers created Google and found the fastest way to index content. When a developer created Facebook, he wanted to get students connected (and also to meet girls). Due in part to the freedom and ease offered by the network and the technical knowledge that they (not many) had, developers were able to create a business from some very simple ideas (e.g. search for information, connect people). And that, dear "web developer", was the beginning of the end. The minute that investment funds saw that the solidity of small firms could create profits without large investments (in the initial stages) was when developers became entrepreneurs. And that is when empowerment came true.

That was a mirror for many but was still far from others' reach. However, computer freaks clearly lead their companies according to values that are very different from the existing ones (old-fashioned, greater inertia, bad habits and stagnant organisational routines) or from non-Internet ecosystem related values. And that is where the so called "dark side" of developers-entrepreneurs appears. The ability to reach a barren landscape first allows you to do whatever you like and build roads so that others can arrive, although they are obliged to take your route.

There was no jQuery when I studied programming 10–15 years ago.[12] It was JavaScript, full stop, and you could make animations, but looking back they now seem very flimsy. It was when there were webs with midi music and gif images that never stopped moving. Smartphones were only seen in films and what really thrilled me was learning ASP (Active Server Pages) and PHP, which I used to request entries and store them in databases. But now it's 2015 and I'm still programming in PHP, over HTML5. Thanks to JavaScript, we now use jQuery to manipulate interfaces and the data are no longer strictly structured so we can use JSON (acronym for JavaScript Object Notation, which is a light format for data exchange) and save the entries in a database like MongoDB. I start swearing when I have to work with Less or SaaS, for instance, but now looking back, I see that these new techniques (i.e. workflows) are nothing more than evolved concepts that we all knew and recognised.

For developers, the biggest evolutionary leap forward may not have arrived with the emergence of a new programming language, but with a language like HTML. In its progressive evolution to HTML5, its capacity has been enlarged and at the same time so has that of other languages, making it possible to advance together and take data manipulation to another level. That ability to make the other languages grow may be what I like most. It's a bit like a midfielder who makes incredible passes to the forwards or that point guard who runs the game and controls the tempo in a basketball match.

It's funny, in spite of several innovations in our developer world, not a one has been capable of creating a real disruption. And what I find even stranger is that thanks to these languages, these incremental innovations, new organisational and business models have actually emerged and have checkmated traditional concepts and business models. Don't forget that the term disruption, which is characteristic, although not exclusive to, dynamic, turbulent and high velocity environments (Eisenhardt 1989), involves a radical break (i.e. a paradigm shift) in a process of constant, progressive and gradual evolution.

Do developers think that there is a disruption that is the same or equivalent to what the big Internet companies are creating in other scopes? That was the initial

---

[12]http://jquery.com/ (last access 12th October 2015).

research question we were going to try to answer in this chapter. My conclusion is that there is not one. Current changes, the speed of such changes, new development patterns, etc., continue without being radically innovative. Developers adapt to the needs of the environment, which makes it a more Darwin-type movement than epigenetic. The following are needed for these dynamics to appear and consolidate:

- Sustained financial capacity from developers or start-ups that want to enter into such dynamics.
- The possibility of attaining certain amounts of intellectual property (patents, copyrights, creative common licences, free software, etc.).
- Access and preparation of their own "human capital" that enable them to penetrate and improve in certain specialist fields or areas of knowledge, such as, for instance, provision of engineering teams or legal counsel needed to defend their positions.
- Marketing research, as well as growth of potential users in different fields that the business group can target (i.e. dominant vectors). In this case, we would also have to consider competition from other business groups (i.e. GAFA, developers and start-ups) to compete on potential markets.

I believe that the epigenetic approach (i.e. EED) to the study of the dynamics observed in the Internet ecosystem makes it possible to clear up the dark side, which at the present time is being played by the same companies that have been capable of true innovation in a disruptive manner in human communication and in the Internet industry. This was, of course, the reason why Gómez-Uranga et al. (2014) introduced the concept of Epigenetic Economic Dynamics. However, it doesn't seem to be the most suitable method to study developers' dynamics (see Sect. 6).

When answering the question of which Internet companies are the most innovative or the biggest ground-breakers as per new ideas, I personally feel that IBM and Oracle, for example, have been much more innovative than the rest of today's GAFA for many years, and having seen the change coming, have been able to defend their market shares. Although I would have to point out that these firms are in sectors that are not as highly visible for most mortals. In spite of that, and not having economies of scale or scalability like GAFA (these phenomena are almost unheard of), these firms managed to revolutionise the world in which we live. Now GAFA are the ones trying to make us believe, almost compulsively, that there are only a couple of development models with a sole approach to making applications, and even that they are only for the goals set by their own marketing gurus. Let me explain, Google, Microsoft and Apple, whose development capacity has enabled them to create the technology base for applications development for their terminals, also oblige us to comply with their specifications (e.g. Application Programming Interface—API). Some are stricter than others, but you have to meet their "terms". Why this unwarranted attack on marketing agents? Because their aim as technology bases is to have their own area for profit-making activities and the marketing agents will sell us the good points but will hide the other part they don't want us to see (e.g. use of personal data for third parties, etc.).

And what was the catalyst that caused people not to give a toss about their privacy and rush off to give their personal information to GAFA? The popularity of mobile terminals (remember mobility was changing everything). Mobile terminals have stirred up everything... business and people. Especially people, since the majority of applications look to entertain us or give us information (and naturally, capture our personal information), so people are ultimately the main target of all these applications and the focus of online services. So that we can get an abstract vision that we can understand, close your eyes, well no, on second thought, don't. You can't read like that so try dreaming while staying awake: we have an application on our terminal that visualises information on public transport. As we have seen earlier, there is no data on our terminal; it comes from server X and is interpreted by the app we have started up. This same application has an interface that works via a browser. So when our customer is in the development phase, he wants to be in the top search position. Google, for instance, has some guidelines that you have to comply with so that its online services place you in a high search position. This is the "for me or against me", which has always existed. However, we don't seem to mind too much because it's abstract. Although we then enter keywords to place our webpage in a top search position, when we access other webpages, we will see the advertising for goods or services we have previously searched. So now think about your privacy and what you'd like it to be in the present or in the future before you click on that dense text.

This is thanks to APIs (Application Programming Interface) which are the services we discussed earlier and which act as brains that prompt us to connect to them and offer us the capacity to work with their information. One of the latest reports by the market research company ComScore (2014) reveals how all the measurements for terminal apps are shooting up while desktop apps indicators remain stable. Think that having the terminal so near us, with our human obsession to look at the mobile terminal, makes developers, or more specifically, companies direct their products to these devices.

Access to the Internet, exposure of our information, granting our privacy so that our information is exploited globally, has been the newest secrets inflating the bubble. This time, I think they are being controlled by investment funds rather than the laws of the states where they are applied, with the aim of avoiding another bankruptcy like the dot.com one that occurred at the start of the new millennium. Now investors know what this is all about (i.e. often regarded as smart capital— Wriston 1998; Sorensen 2007).

A documentary on epigenetics that I revised a couple of years ago, when one of the editors of the book began to talk about his work on epigenetics and the possibility of writing a chapter for a book on EED, so that I could get a good grasp of the biology part, explained how a grandmother's terrible trauma was reflected in her grandchildren. The kids were exposed to a context similar to the one their grandmother experienced, which had a tremendously negative influence on her, and they automatically became stressed more quickly. We can understand that the very experience, the technological and regulatory chaos we live in, has brought about a change in investors' DNA, ensuring that the digital business model of exploitation and investment does not lead us to another disaster like the one in 2001.

5 The Future: The Evolution of Developers

What will developers be doing in 10 or 20 years? With a bit of luck, the same thing they are today, but adapted to a new environment (i.e. context), and if this good fortune bypasses them, they will be doing exactly what they are today unless they are capable of evolving. As I mentioned before, there are a lot of programmers and GAFA are creating APIs to work with their information and capacity to exploit it. So, on the one hand, we have top level developers focusing on standards development to serve as the base for secondary level development, such as that generated by apps. There's nothing bad about this a priori, but development is provided by big companies (i.e. GAFA) which are the ones arriving with new procedures that do no more than "build" the play area (i.e. we are referring to the APIs developed by the large Internet players). So will the future be controlled by a few big firms? Definitely. Particularly, and this where we get into one of those great conversations, when we take into account that the free code concept is possibly used more than ever by big companies. Just think that companies like Apple (but we could say any GAFA included in this book) create products that have such tremendous social and economic impact, for instance on their followers' image, and can even create social division between followers and non-followers. This prompts developers to adapt to their workflows, tools and hardware, and it is not with the aim of taking full advantage of third party applications, services or developers. It's the hardware they propose and above all, their operating systems that coordinate physical devices, the cloud and in general, all their users' public and private information.

We often find caricatures of these mega firms in series and films although if we take a closer look, their software has changed from being proprietary to being free. The Internet ecosystem companies have fought like wild beasts for patents and standards to keep their competitors in check and develop that same software. Thanks to free software, new business models can be developed and foster the creation of new markets.

We can enter "Web 3.0" in a browser and almost all the responses we find will have a common denominator, "data order". Well, for various reasons, I just don't like labels so I am going to try to rewind to understand the current state of what we now see every day. Did a Web 2.0 exist? I don't think so. Just because it occurred to some people (sorry, Tim O'Reilly)[13] to call something by a certain name doesn't mean it was true. Hundreds, even thousands of designers rushed out to copy what was understood as the erroneously termed "2.0 style". I saw marketers sell 2.0 projects which were 100 % the same, and the only new feature was a social media site plugin.

Of course, as a concept to mark an evolutionary point in history, it sounded good. But none of us are able to think that on Tuesday, 5 October 2004, the Web 2.0 started up. I think a lot of people would think "You are wrong, the Web 2.0 was socialisation, blah blah blah…". Fair enough. But was it like that before social media sites or was it the other way round? Did the concept come from the popularity and use of a certain type of projects?

---

[13]See: http://www.oreilly.com/tim/bio.html (last access 12th October 2015).

Hoping that the same doesn't happen with the Web 3.0, it is defined by the World Wide Web Consortium (W3C) as the Semantic Web.[14] And we are already making strides in that sense. If not, what is Big Data and why are we making so much noise about it? It is the first step in the above mentioned "data order". We are moving to an Internet of data or, in other words, massive information. These are the main characteristics of the Web 3.0[15]:

- *Intelligence. The Semantic Web project known as Web 3.0 intends to create a method to classify Internet pages, a tagging system that not only allows browsers to find the information on the network but to understand it. By achieving this objective, users can access the Web to ask in their language, the Web will understand that language, and learn the result of the searches for the next operations. Although the fact of learning is just to save values and apply statistics to them; the change in the labelling is the important thing here.*
- *Sociability. Social communities become more exclusive and complex. Social media sites increase as well as the ways in which they connect to their members. It begins to be considered normal for a person to have several identities in their virtual life and the possibility of migrating the identity from one network to another. I would like to remind that social networks are private companies. Are we going to continue giving our data to private companies? This increasingly resembles a dystopian film. We should learn, as discussed before, to apply the digital pedagogy and prevent abuses as we do in our physical life.*
- *Speed. Video broadcasts on the network and the creation of portals devoted to this task, such as YouTube, are possible thanks to fast user connections. The main telecommunications operators have started to implement fibre optics for users with wideband connections up to 3 Mbps ADSL which would convert to speeds from 30Mbps to 1000 Mbps or even faster.*
- *Open. Free software, standards and Creative Commons licences have become commonplace on the Internet. Information is freely distributed on the Web, preventing sole ownership. Capital gains on information are discontinued in favour of more democratic use.*
- *Ubiquitousness. Personal computers will become obsolete due to the multifunctional nature of mobile phone and other portable devices. With the arrival of email on BlackBerry phones on desktops, Apple and iPhone are expected to include the Web. Small screens get bigger and higher resolution, enabling better visualisation of web content. The range of wireless networks and last generation phones increases, expanding network coverage.*
- *User-friendliness. Internet users that visit a new website have to devote a certain amount of time to learning how to use it. The new design tendencies look for standards for a Web with more homogenous and more easily recognisable functions, besides creating spaces that users can set up however they like.*

---

[14]See: http://www.w3.org/standards/semanticweb/ (last access 12th October 2015).

[15]See: http://datateca.unad.edu.co/contenidos/MDL000/ContenidoTelematica/caractersticas_de_la_web_30.html (last access 12th October 2015).

- **Distribution**. *Programmes and information become small pieces distributed by the Web and are capable of working together. Internet users can collect and mix these pieces to carry out certain actions. The Web thus becomes an enormous space that can run like a universal computer. Distributed computation systems—systems which connect the power of many computers in one entity—become a commonplace option of operating systems.*
- **Tridimensionality**. *Tridimensional spaces in the form of virtual worlds as game and online courses will become increasingly common. There will be new devices to move around the Web, different from keyboards, the mouse and optic pencils.*

However, I find that the drawbacks, obstacles or difficulties that have to be overcome for its successful implementation are very serious and specific[16]:

- *The decentralisation of Web management offers developers the freedom to freely create tags and the ontologies they need to make their webpages sensible. However, the downside of this freedom is that various developers could use different tags at the same time to describe the same things. This could enormously complicate the comparisons for the machines due to the possible ambiguity of terms to refer to the same thing.*
- *There is criticism about what is philosophically known as the "identity problem", which centres (in its computational transposition into the Semantic Web) on whether an internationalized resource identifier (IRI) only represents the web resource that it makes reference to, or, in contrast, the implicit concept in the referenced web resource. (e.g. The IRI shows the path of the webpage of an institution. Does it really represent the institution in itself or just a webpage written about it?). The point is important when establishing trustworthy sources or resources considered "axioms" from which knowledge taken can be inferred to be true.*
- *Finally, the biggest obstacle of all is the Semantic Web's dependence on establishment of adequate ontologies and rules to give it meaning. Building ontologies requires a great deal of work and is actually the central issue and where most of the work to build the Semantic Web is done. Will companies and people be capable of devoting the necessary time and resources to creating adequate ontologies so the existing websites can "understand" the Semantic Web? Will their ontologies maintain and evolve as the content of their websites change?*
- *Some sceptical developers disagree with the approach that the Semantic Web should be totally dependent on establishing ontologies and rules, to the point of arguing that the project is unfeasible because of its huge dimension. They contend that the task of creating and maintaining such complex descriptive files is too much work for most people and furthermore, that companies are not likely to devote the necessary time and resources and add the necessary metadata to the existing websites so that they can work properly on the Semantic Web.*

So we see identity problems, creative freedom, possible personal rights at the digital level, etc., are basically the same problems we have today, but in another

---

[16]See: https://sites.google.com/site/groupccygv/wiki-del-proyecto/web-2-0/hacia-la-web-3-0-la-web-semantica (last access 12th October 2015).

similar context, not very different from our reality. I fear that as it happens at present times, there will be a consortium of actors that recommend certain actions, patterns, etc., but without getting in the activities of the dominant players. Instead of leading the progress, they may simply serve as a collector ideas and developments and provide a forum for standardisation. I'd like our readers to be aware that we are on the verge of changes that could go one way or the other, leading to very different things. Gartner (2014) stated that in 2014, 73 % of the people interviewed were going to invest in Big Data projects in the following years (in 2013 it was 63 %), so we are at the gateway of new dilemmas about how information is handled on the Internet. In this respect, Google, for instance, is doing its homework and now has tools to process large amounts of data very quickly, which gives it the capacity to provide them to generate new applications for third party firms and thus extend the field to the following source of data, which is things. And the rest of GAFA are doing the same thing.

6 Making Big Data Known

So what does Big Data mean at the developer level? Actually, not very much for a technician working as a front-developer in a medium-sized firm. There is something alarming, though. Right now, code developers have access to almost anything to run a project, a database, frameworks, space to place files on a server. However, as of now, access to data organisation is not going to be free. It will be on order from a big Internet firm. And this will divide companies that work directly with data from those that have to rent them, instead of their being freely accessible to anybody.

So we'll be facing the same contradictions we have today (i.e. Does Google manipulate the data it displays in searches?), with all the questions that keep coming up. Everything will continue to be the same until what I foresee as a new turning point, which could be when "things" send data directly to our services.

Lohr (2015) points out how IBM, the hardware and software manufacturer, announced that a considerable sum of money was going to be put into Big Data. More specifically, into the Big Data free code software, Apache Spark. That will trigger another cycle of data explosion, communication, interaction and we will have a 3–4-year period in which to form part of this new data flow, with communication between objects emerging soon afterwards. This will make man more passive than ever as per socialisation and interaction with other human beings.

And what will developers do then? If we are active, we should learn to manipulate those data, since many, although not all, future companies will have a business model based on this type of services. So becoming a provider of these services could be very profitable.

Imagine what the HTML5 standards and changes have meant to some. It has allowed the creation of a large ecosystem, an entire landscape, down to the very last detail. Then the development teams went in and suggested putting in a pipe network and began to change the way data were handled. They made non-relational (i.e. NoSQL) databases popular and generated frameworks to handle them, thanks to their strengths and new improvements. Let's not forget that the world contains an increasing amount of software, whereas hardware is shifting from an inert object to become a semi-intelligent being, due to the intangibility of software. This is starting to open a gap between people who are capable of understanding, interacting, working and negotiating with intangible information and those who can't. In just a few years, the rift will exist between people who have a close relationship with intelligent objects and those who don't. And when the time comes, we'll see how society is fragmented and what social groups are formed. Just give it time.

Going beyond this, it's plausible to think that the Big Data system itself will become an axis of the Internet. To offer a simile, Big Data services will play a similar role to today's browsers, but perhaps at a more concealed level from the final user but which developers will be required to know about. Who knows? A couple of young people who avoid investment funds might come along and turn out to be capable of coming up with a new focus for data use and create a totally alternative business model. They might become a major Internet company and force developers to adapt and evolve with them. Anything is possible.

And on top of that, revising the Web 4.0 guidelines (yes! Web 4.0!) people like Nova Spivack say that around 2020–2033, network intelligence will take a huge leap forward, which is the inspiration for Web 4.0 and it will be similar to human reasoning. Like today, we will make the sum of many services be regarded as a standard. The question, as I mentioned earlier, depends on whether it's going to be an open model or one by request run by the big firms that exist then. Depending on this, developers should learn to use new frameworks and methodologies to run what our customers want or what firms demand, in the event that we create a start-up. Whatever happens, I don't think it will very different from today's reality, but probably from another environment or context. There will still be top level developers and others who are pulled along by the tools or language evolution developed by the former.

So, after all that we have talked about, I return to the question: is it disruptive? I don't think so, but what we do see increasingly is a pattern and, in our case, GAFA are like a sun that erupts from time to time, resulting in a true disruption. And, as I said, it is becoming a pattern and eruptions are fantastic to look at but are extremely dangerous.

7 How Can We Study the Evolution of Developers? an Analytics Proposal Based on Quantum Physics

In the previous chapters (mainly Chapter "Introducing an Epigenetic Approach for the Study of Internet Industry Groups" by Gómez Uranga et al.) we extensively developed the EED model which was later applied to GAFA. However, in the third section of this chapter, we reached the conclusion that the developers' dynamics seem to fit Darwinist characteristics more than epigenetic. As a result, although the analytical framework of the EED does adapt to the study of the big Internet business groups' dynamics in an interesting manner, we don't believe this happens with the other side of the coin, which is developer dynamics. This, in turn, leads us to think that, in view of the characteristics of the Internet ecosystem and the size of the population which is the object of study (i.e. 18.5 million developers around the world) perhaps we need to think of an analytical framework that will enable us to understand the phenomena that occur in the world of developers.

In this section, we intend to introduce a new analogy which is related to quantum physics to apply it to the case of developers and attain a more robust analytical, conceptual and empirical understanding of their characteristics and dynamics. Of course, this is only a methodological proposal which must be strengthened and applied, so there remain many stages that must first be studied and many challenges to be overcome with this 'cross-fertilization'. First, bringing in new concepts implies the need to develop new analytical approaches. Second, these conceptual approaches need to be translated into methodological tools. Third, in order to validate these new approaches, it is necessary to gather data from the different actors that are operating and shaping the Internet ecosystem (i.e. developers and start-ups). Gathering these data is a task in itself. There are millions of start-up companies constituted by developers, scattered across the globe and which are very small in size. Following them constitutes a difficulty itself as their traces are not observed in the market from the moment of constitution, but rather when they are acquired by large players. In this respect, we will develop what fundamentals/properties of this latter discipline could be imported to our fields of study. We therefore feel that the application of epigenetics to study the behaviour of the big Internet business groups through EED could be supplemented with an appropriate use of a quantum approach for the case of developers.

As noted, in order to understand the dynamics of the Internet ecosystem, it is necessary to know the dynamics of developers, which materialise in the creation of new technology start-ups. Developers (i.e. entrepreneurs) are key in explaining epigenetic dynamics. Examples of GAFA absorbing entrepreneurial ventures include Facebook acquiring Instagram, Whattssap or Oculus; Google acquiring Nest; or Microsoft acquiring Mojang for instance (see Chapter "Epigenetic Economics Dynamics in the Internet Ecosystem" by Zabala-Iturriagagoitia et al.).

Developers are becoming the cornerstone of the Internet's rapid development and the abrupt growth of the large industry groups dominating it. The literature increasingly emphasises the relevance of entrepreneurs in employment generation (Autio et al. 2014; Bruton et al. 2013; Engelen et al. 2014; Mazzucato 2011). However, in spite of the key role developers play in the dynamics of the Internet ecosystem, there is hardly any evidence regarding which dominant vectors are guiding developer activity, their economic impact both in terms of employment generation in the geographical areas where they are located, the generation of added value, the challenges they encounter when facing competition from GAFA, or their strategies in relation to intellectual property protection. We believe the principles of quantum physics could be adopted to systematically explain the evolution and changes in the orientation of developers and how they are influenced and, at the same time, affect the changes in the Internet ecosystem.

Quantum physics is characterized by three principles: quantum superposition, entanglement and collapse.

- Quantum superposition: Schrödinger's uncertainty principle, determines that a particle is in all the states it could potentially be in. This is illustrated by the metaphor of Schrödinger's cat, which is alive and dead at the same time (i.e. a particle staying in all possible states at the same time).
- Quantum entanglement: this principle shows how a set of particles cannot be defined as single particles with defined states, but rather as a system with a single wave function. The strong relationships between the particles (entanglement) make the measurements done on a system appear to instantly have an influence on other systems that the original system is intertwined with, no matter what the separation among them is. In other words, a particle affects the system as a whole. Accordingly, the distribution of probability of the particle being located in a concrete state is dependent on the system as a whole.
- Quantum collapse: refers to the transition of a quantum system from a superposition of states to a concrete state. It is related to the quantum superposition principle, inasmuch as a particle, which can potentially be in any possible state, when making an observation on it, will collapse to a concrete state with a defined value. The process is also known as collapse of the wave function or collapse of quantum states, and the probability of collapsing to a given state is determined by the wave function of the system before the collapse.

The relationship established between the dynamics of these three quantum principles and developers is as follows (Table 2). Initially, given the horizontal character of the software industry and the generic capabilities required in it, every developer could be oriented to all potential activities and industries (i.e. quantum superposition). However, developers opt for certain dominant vectors (Suárez et al. 2015). This decision to focus on certain markets would be equivalent to the quantum collapse. As a result, the introduction of developers into certain markets alters the situation in which that market or sector showed during a previous state (quantum entanglement).

Application developers are becoming increasingly important, not only for the dynamics of the Internet ecosystem, but also for new employment creation and economic growth. The dynamism of current societies is based on the development of applications and on entrepreneurship to a greater extent (Glassdoor 2015; Newbert et al. 2008). Developers have multiple directions or dominant vectors they orient toward. Depending on their location and the characteristics of the environment in each location, they will collapse into these vectors with different probabilities. The key lies in identifying the dominant vector that may guide the activities of entrepreneurs (i.e. developers) and which rely on the higher efficiency in each location (Zabala-Iturriagagoitia et al. 2007). Finding out which the dominant vectors are in each territory, policies (e.g. entrepreneurship, innovation,

Table 2 Relationship between the quantum properties and developer dynamics

	Millions of developers	Billions of electrons and particles
Quantum superposition	A developer can orient to all the potential states it could potentially be in	A particle is in all the states it could potentially be in
Quantum entanglement	Developers depend on global relationships and requirements	Particles lose their meaning as isolated elements
	New firm entry (created by new developers) has an immediate influence on the ecosystems, altering their previous situation	They are precisely defined by their entanglement with other particles
Quantum collapse	When making an observation on the start-up, this will be specifically defined in a particular state from all the existing options it could initially be oriented to	When observing a given particle or element, a perfectly defined state is created

Source Own elaboration

employment, education, regulatory, tax, etc.) may imply greater effectiveness when supporting these entrepreneurs' activities.

Naturally, as in all entrepreneurial processes, a large share of the new entrants fails. This is where the quantum probability becomes important. In the stage prior to entrepreneurship, there is superposition of states since the entrepreneurial firm can both fail and survive. Therefore, the quantum analysis would be equivalent to a probabilistic analysis. The novelty of the project lies in that there are millions of developers, but a minority of these succeeds and becomes firms of a certain size and reaches a certain degree of success on the market. It is here that the quantum approach meets EED, as the difficulties entrepreneurs face are, to a great extent, due to the dynamics of GAFA. A large share of the developers has great interest in being acquired by GAFA, since they know they cannot outcompete them due to their financial power. What is more, these large groups often even "own" their developers (e.g. through the organisation of huge contests or hackathons). As we have described in this chapter, GAFA often act as "lodestars" that guide the action of the developers themselves. That is why we have often referred to the Internet ecosystem as the two sides of the same coin (Perks et al. 2012).

In order to study the dynamics of developers and their start-ups, it would be possible to rely on the use of quantum Bayesianism. Quantum Bayesianism was introduced in 2002 by Caves et al. (2002), unifying quantum physics with probabilities. The adjective Bayesian is due to Bayes' theorem and the conditional probabilities used in inference processes. In Bayes' theorem, evidence (or observation) is used to infer the probability that a hypothesis may be true. The basic idea behind Bayesian probability is therefore a calculation of the consistency of the credibility of a certain hypothesis. In the Bayesian context, the term "degree of belief" is used, since the expert believes that something can be real (i.e. can take place) with a certain level of belief, and therefore cannot assert whether something is true or not. The actual beliefs come from external sources, and the researcher sets the degree to which something can happen by assigning an ex-ante probability. As the system is collapsed according to different experiments, the researcher defines the probabilities for the different options to actually take place according to the observations of those experiments (i.e. the results of each collapse, which will be different).

Certain environmental variables may be significant in assigning these ex-ante probabilities, which may be defined by the researcher. A starting point for this definition can be Jeffreys' models (1961, 1973). Using regression models in which the target variable (i.e. the most efficient dominant vectors in each location) is a probability function, and in which variables related to the environment where the developers are located are used as explanatory variables, it is possible to reassign the initial ex-ante probabilities using Bayes' theorem. Consequently, the new probabilities are defined, which modify the initial beliefs according to the information provided by the data. The model will be amended as new variables related to the environments are introduced, obtaining the Bayes' factor, as a result of the likelihood ratio of the first and second models once the new variables have been included. This results in a series of nested models that seek efficiency in the probabilities of the decisions to be made by the developers concerning their activities.

The previous methodology requires identifying a set of systemic variables representing each environment. Some of these contextual variables might be:

- Level A: Quantitative indicators

  - General structural indicators such as those included in the Innovation Union Scoreboard.
  - Other relevant indicators related to entrepreneurship such as youth unemployment, new business creation, survival rates, sectors with higher growth rates, etc.
  - Most relevant economic sectors in each country.
  - Type of firms according to age, owner, size, R&D investments.
  - Availability of a trained labour force: share of the population with higher education, disciplines in which people in the country are more specialised.
  - Availability of venture, seed and risk capital.
  - Extent to which the Internet is implemented in the country, share of purchasing over the Internet.

- Level B: Qualitative indicators

  - Vertical priorities in terms the different governments' policy such as health, sustainability, manufacturing, energy, creative industries, etc.
  - Public support for entrepreneurial action: are there entrepreneurship policies? Are they focused on specific industries?
  - Presence or absence of large multinational corporations that may act as 'drivers' of their respective economies, which may attract not only developers and their start-ups but also other large corporations to the 'hot spots' where they are located.
  - Quality of the institutions.
  - Comprehensiveness and level of relationships among the different parts of the innovation system.

- Entrepreneur-friendly climate.
- Share of people with programming skills (coding, Big Data, cloud computing, mobile, data visualisation, user experience designers).

These potential variables are far from being comprehensive and this list should be considered a preliminary approximation. One of the weaknesses of this model is that there are multiple variables that cannot be included, such as the developer's subjectivity (e.g. whether its motivation is to grow or not to grow and just create a small amount of employment), the team size (particularly when the firm is constituted), family background, level of competence and skills, etc. These variables require other non-economic sciences such as genetics, sociology or political science, among others. The model also needs to be dynamic in order to capture the evolution in the different territories. This requires constant updating of the data, year by year. As indicated, the main challenge involved in the operationalization of this quantum approach is related to data acquisition, since many of the potential indicators listed above are not systematically collected. Therefore, quantitative and qualitative methods need to be combined to gather the necessary data. Finally, in order to assign ex-ante probabilities to the direction that developers may take, it is necessary to assign some weights to the chosen indicators, which also raises certain challenges as the weight of the structural indicators will vary from territory to territory.

The interest and potential usefulness of the quantum approach is manifold. First, it can be useful for policy makers to shape their policies, priorities, financial investments and the alignment among policy domains (e.g. environment, health, education, etc.). It also points out the environmental elements that need to be improved, supported, included or even eliminated so developers in their respective territories can have higher probabilities of success. Third, it can also assist GAFA in their diversification strategies, as it captures the dominant vectors on the one hand, and the type of vector that developers in different territories orient to. Fourth, the model can also be useful for the developers themselves, so as to know which locations are the most efficient, depending on the sector they want to focus on. Finally, the model can be of great interest for investors, enabling them to know where developers are, according to the sector they are interested in.

The previous quantitative and qualitative variables should embrace both the local and the global levels. Naturally, many of these variables are local in character. However, many others are global, not only because the dynamics of the GAFA are global, but also because the big venture capital firms, the lines of action (which are often defined at the European level) and even the Internet market are global. This does not preclude that local analyses of certain geographical contexts can also be carried out. With the very initial model we would be drawing thick lines, which identify the efficient ones? at the national level (a dominant vector for each country). However, ideally, the more variables at the local scale that could be included, the more accurate the dynamics of entrepreneurs and developers could be. It would be equivalent to zooming into see how the thick lines at the national level are also divided into thinner lines. However, this is very much dependent on the challenges involved in data acquisition and the level at which these can be gathered.

References

Autio, E., Kenney, M., Mustar, P., Siegel, D., & Wrights, M. (2014). Entrepreneurial innovation: The importance of context. *Research Policy, 43*, 1097–1108.

Bruton, G. D., Ketchen, D. J, Jr, & Ireland, R. D. (2013). Entrepreneurship as a solution to poverty. *Journal of Business Venturing, 28*, 683–689.

Caves, C. M., Fuchs, C. A., & Schack, R. (2002). Quantum probabilities as Bayesian probabilities. *Physical Review A, 65*(2), 022305.

Comscore (2014). The U.S. mobile app report.

Eisenhardt, K. M. (1989). Making fast strategic decisions in high-velocity environments. *The Academy of Management Journal, 32*(3), 543–576.

Engelen, A., Kube, H., Schmidt, S., & Flatten, T. C. (2014). Entrepreneurial orientation in turbulent environments: The moderating role of absorptive capacity. *Research Policy, 43*, 1353–1369.

Gartner (2014). Survey Analysis: Big Data investment grows but deployments remain scarce in 2014. http://www.gartner.com/newsroom/id/2848718. Accessed 19 Oct 2015.

Glassdoor (2015). http://www.glassdoor.com/blog/jobs-america/. Accessed 19 Oct 2015.

Gómez-Uranga, M., Miguel, J. C., & Zabala-Iturriagagoitia, J. M. (2014). Epigenetic economic dynamics: The evolution of big internet business ecosystems, evidence for patents. *Technovation, 34*(3), 177–189.

IDC (2013). 2014 worldwide software developer and ict-skilled worker estimates. International Data Corporation.

Jeffreys, H. (1961). *Theory of probability*. New York: Oxford University Press.

Jeffreys, H. (1973). *Scientific inference*. Cambridge, England: Cambridge University Press.

Kuhn, T. S. (1962). *The structure of scientific revolutions*. Chicago: University of Chicago Press.

Lohr, S. (2015). IBM invests to help open-source big data software and itself. http://bits.blogs.nytimes.com/2015/06/15/ibm-invests-to-help-open-source-big-data-software-and-itself/?_r=0. Accessed 19 Oct 2015.

Mazzucato, M. (2011). *The entrepreneurial state*. London: Demos.

Moore, J. F. (1993). Predators and prey: A new ecology of competition. *Harvard Business Review, 71*(3), 75–86.

Newbert, S. L., Gopalakrishnan, S., & Kirchhoff, B. A. (2008). Looking beyond resources: Exploring the importance of entrepreneurship to firm-level competitive advantage in technologically intensive industries. *Technovation, 28*, 6–19.

Perks, H., Gruber, T., & Edvardsson, B. (2012). Co-creation in radical service innovation: a systematic analysis of microlevel processes. *Journal of Product Innovation Management, 29*(6), 935–951.

Quintero, C. (2015). Who invests in hardware startups? http://techcrunch.com/2015/09/12/who-invests-in-hardware-startups/. Accessed 19 Oct 2015.

Sorensen, M. (2007). How smart is smart money? A two-sided matching model of venture capital. *The Journal of Finance, 62*(6), 2725–2762.

Suárez, F. F., Grodal, S., & Gotsopoulos, A. (2015). Perfect timing? Dominant category, dominant design, and the window of opportunity for firm entry. *Strategic Management Journal, 36*, 437–448.

Veracode (2013). The history of programming languages. infographic. https://www.veracode.com/blog/2013/04/the-history-of-programming-languages-infographic. Accessed 19 Oct 2015.

Wriston, W. B. (1998). Dumb networks and smart capital. *Cato Journal, 17*(3), 333–340.

Zabala-Iturriagagoitia, J. M., Voigt, P., Gutiérrez-Gracia, A., & Jiménez-Sáez, F. (2007). Regional innovation systems: How to assess performance. *Regional Studies, 41*(5), 661–672.

Future Paths of Evolution in the Digital Ecosystem

Jorge Emiliano Pérez Martínez and Silvia Serrano Calle

1 Introduction

The term digital ecosystem was coined by Fransman (2010), who understood it as all the activities created around the technological development of the Internet, ranging from network infrastructures to applications and end-user services. Since its origins, the digital ecosystem has undergone expansion and growth of new business models, social relationships, and exchange of information which were previously unheard of. Many of the new economic agents that have emerged as a result of new opportunities brought by network development have broken with the conventional model of business growth and evolution. We are witnessing a relatively recent process, which particularly began in the 1990s. The process shows one particularly relevant feature; it is a system that evolves very rapidly and in which conventional order and power patterns have changed significantly.

It was in the 1960s that the US Department of Defence began to develop the Arpanet project. The research project was originally a communications network to connect university computer groups to share computer resources that were only available at certain research centres. The development of the Internet architecture and the new communication protocols, such as TCP/IP created by Cerf and Kahn (vid. Cerf and Kahn, 1974), that were later improved (vid. Leiner et al. 1985), to create the World Wide Web (another work of individual genius by Tim Berners-Lee at CERN) in 1990, was a complex process.

J.E. Pérez Martínez · S. Serrano Calle (✉)
Escuela Técnica Superior de Ingenieros de Telecomunicación (ETSIT),
Department of Signals, Systems and Radiocommunications,
Technical University of Madrid (UPM), Madrid, Spain
e-mail: Silvia.serrano@upm.es

J.E. Pérez Martínez
e-mail: Jorge.perez.martinez@upm.es

© Springer International Publishing Switzerland 2016
M. Gómez-Uranga et al. (eds.), *Dynamics of Big Internet Industry Groups and Future Trends*, DOI 10.1007/978-3-319-31147-0_6

The design of the Internet as an open shared network with standards and open protocols, and a structure designed to be scalable, where new nodes could be added without altering its essence (Leiner et al. 1997), has proved to be a technological and economic success. The Internet has been a wonderful laboratory for research, fostering communication and innovation. Open collaboration has strengthened the power of the network in this virtuous circle although the big leap toward the digital ecosystem as we know it today was boosted by two key aspects. Without them, the Internet would have merely remained an interesting academic experience and never have become a social and economic phenomenon of paramount importance. These aspects were the expansion of telecommunications infrastructures and the availability of the resources to guarantee permanent connectivity anywhere, the ubiquity of the network, and also the necessary development of reasonably priced user-friendly terminals, the user interfaces with connection capacity and access to the Internet. Internet offers a new space to host the information, providing new services and contents demanded by users. The Internet has caused disruption in numerous scopes. It is not only a technological advance, but also an economic one, as it alters the conventional relationships established in the classic supply and demand model, and also what could be considered a more subversive element, that of human relationships. This new immediate global connection model has changed social relationships, from the closest family nucleus to relations at work, and brought a new political and social order.

There is a great deal of literature from the field of economics on different theories that help to explain the changes driven by technology development (see Chapter "Introducing an Epigenetic Approach for the Study of Internet Industry Groups" in this book by Gómez-Uranga et al.). The development process of economic and scientific models shares many similarities with it. The paradigm of technological progress understood from the conventional approach, as pointed out by Kuhn (1962), determines not only scientific-technological research but also innovative thinking. As Kuhn stated, the evolution of ideas in the scope of science traditionally occurs due to the accumulation of discoveries which are largely works of individual genius. When there is an exception to the model that significantly prevents it from fitting in the traditional paradigm, a disruption is created that will lead to a change in the system. The new paradigm enables explanation of the disruption in an orderly manner. In this sense, the arrival of the Internet and the enormous development in telecommunication technologies in recent decades have allowed the creation of a new ecosystem: the digital one whose rapid development we are witnessing. The new approach we are introducing in this book, Epigenetic Economic Dynamics (EED), enables us to explain the dynamics occurring in the digital ecosystem from a disruptive perspective as regards conventional models. The new ecosystem is distinctive for carrying a strong innovation component in various scopes: technology, economic and social. Another feature of the new model represented by the Internet is the speed at which the ecosystem itself evolves. This may be analysed in terms of a new disruptive phenomenon which a new paradigm such as the epigenetic one can reasonably explain.

The environment of the so-called digital ecosystem shows some particular characteristics that mark a change in comparison to traditional innovation models and the frameworks of the paradigms they are set within. These characteristics have determined their development and will affect their evolution in future. Some of the most outstanding of them are: (i) It is a global phenomenon; (ii) It has an open architecture; (iii) It embraces high creativity and innovation levels; (iv) It changes and adopts innovations quickly; (v) It eases user interactivity with technologies and applications and (vi) It produces economies of scale and scope.

The firms in this ecosystem operate in an environment which offers constant opportunities but demands ongoing efforts to keep pace with the ecosystem's growth and capture the value it is continuously generating. The concept of EED makes it possible to analyse the dynamics and processes that companies are subjected to as the ecosystem evolves.

2 Technological Progress and Expansion of the Digital Ecosystem

Technological evolution occurs at different levels in the ecosystem. The Internet ecosystem as we know it today would not have been possible without the network infrastructures, the operability of critical Internet resources and the backing of governments, particularly that of the U.S., which took an active role in the Internet's evolution and development.

The digital ecosystem has evolved in recent decades although, in essence, the Internet has maintained its original open architecture. It is a model which facilitates interactivity of the agents that form the ecosystem. Internet architecture is tier-based with the communication capacity distributed in each node. It transfers basic information units or packets via packet switching technology by using specific protocols developed for the Internet (the TCP/IP model). This is done via multiple communication channels so that access to the information can follow many alternate routes to reach the destination.

One of the recent key phenomena is greater transmission and reception capabilities in mobile services. Driven by users' demands and the availability of equipment and terminals permanently connected to the network, mobile wideband services are the ones showing the highest global growth in recent years. In 2014, mobile data traffic increased by 69 % over the previous year, reaching 2.5 Exabyte per month at the end of 2014 (Cisco 2015). Projections indicate that mobile data traffic on the Internet is expected to continue growing. In 2014, mobile data traffic was nearly 30 times the size of the entire global Internet in 2000. Mobile video traffic is the most demanded service by network users at the present time, with over 497 million new connections and mobile devices (e.g. smartphones, tablets, etc.) added in 2014.

The Internet is a global phenomenon. The total number of global connections in 2014 (number of devices) reached 7.4 billion (Cisco, 2015). The demands of new users accessing the network, plus those who are already connected, mean that higher capacity mobile wideband is needed. Mobile wideband is expanding parallel to the development of new services and applications in the Internet environment. In this new environment, firms operating in different fields of production should work together. The success and expansion of many of the latest innovations and developments (i.e., applications, operating systems, terminals, etc.) is based on their capacity to interact, ensuring that the new products are compatible with existing ones.

Mobile wideband expansion requires a limited resource, or public commodity, which is the radio spectrum. There are several possible technology solutions to guarantee such high connectivity levels without compromising the quality of the service. More bands should be enabled, jointly with other complementary solutions such as re-use of the spectrum resources currently being used, improving spectral efficiency and increasing performance and capacity of the channels per megahertz of the spectrum used, particularly for the applications and services most demanded by users in certain geographic areas. 4G LTE (Long-Term Evolution) technology-based wireless communications and new developments make it possible to improve and increase the network's capacity (see Chapter "4G Technology: The Role of TelecomCarriers" in this book by Araujo and Urizar on this very matter).

The Internet is a connection, information and exchange platform. As the digital ecosystem expands, it includes new services on a daily basis; new connections are being made to devices and connectivity is added as everyday objects integrate technology. The need for better connection and wideband capacity is an indication of future tendencies. High-speed Internet connections currently generate more average traffic on the network than slower ones. This situation seems reasonable insofar as 4G users report higher quality and satisfaction with their mobile devices when accessing to web services and online content. Growth in the use of new connection devices is mostly related to smartphones (which only accounted for 29 % of devices in 2014 but 69 % of total handset traffic) and tablets (74 million connections in 2014), vid. Cisco (2015). Globally, there were nearly 109 million wearable devices in 2014 generating 15 petabytes of monthly traffic.

In developed nations, where a higher percentage of the population is connected to the Internet, users normally access the network with more than one mobile device. Mobile Wi-Fi, (also called Mi-Fi) is undergoing dramatic growth. Many smartphones have an integrated Wi-Fi hotspot that other Wi-Fi enabled devices can use to connect to the Internet. These devices may use the same mobile phone number or telecommunications operators may not assign telephone numbers to all the devices that connect via their networks. This is the case of devices which are not designed for voice communication, but only data transfer, such as e-readers, tablets, laptops or modem cards and mobile Wi-Fi hotspots. Use of these devices has increased sharply in recent years and projections predict continued growth in future. Services providers and operators compete against each other and also with other companies (with different genomic instructions), different sales strategies (some of which may be set within the EED paradigm framework such as their genetic footprint), in technical aspects such as speed, coverage and price of the services, or with mixed models that offer users other devices and terminals. Service providers and operators often set up alliances with other business groups (i.e., equipment providers) in the ecosystem and create synergies in groups that have opted for vertical integration of business activities, adding new roles to the DNA, according to the EED approach. Machine-to-machine (M2M) connections are also increasing. M2M systems include transport management systems, vehicle tracking and fleet management, security in homes or businesses, other telematics services or smart grid devices.

Some recent studies point to a disparity between users' demands for mobile wideband and the increases in capacity that technology can actually reach, taking the spectrum's limited nature into account and the current allocation (Clarke, 2014). This imbalance is due to economic factors, infrastructure deployment and improvement costs; oversight related to the changes required in regulations and adoption of new standards; and political aspects related to public policy and social welfare. If the necessary adjustments are not made, the wireless network capacity may hinder expansion in some regions, widening the digital divide between countries.

2.1 Development of New Networks

The digital ecosystem's growth is based on one essential pillar that supports the entire system: infrastructures. This is a vital factor which has fostered and affected the network's expansion and evolution from the beginning. The digital ecosystem should not be limited by geographical boundaries on a scenario like today's where the available existing technology guarantees connectivity. Nevertheless, there is a considerable digital divide between countries. Whereas networks have expanded greatly in developed nations and the coverage ratios for the entire potential population are slightly above 90 % in urban areas (i.e. reaching a potential of 99.7 % in the case of the USA—Federal Communications Commission 2014a), there are still over 4000 million people that have no access to the Internet. Expansion of the digital ecosystem to the entire world population raises a challenge. Nearly 3 billion people had access to the Internet in 2014 and two-thirds of them were living in developed nations. The remaining third is the population of developing nations. Of these user connections to the Internet, 2.3 billion were mobile wideband, although in this case the distribution between users in developed and developing countries was more proportionate. Figures in 2014 were 45 % for the former and 55 % for the latter (ITU, 2015).

Ideal access to the network can be defined as that which allows Internet connection at the appropriate speed to access the desired content when and where the user chooses. From a strictly technical point of view, connectivity is strengthened by the availability of different technologically suitable solutions to achieve global deployment that guarantees connectivity. Furthermore, development will be fostered if the deployment is economically competitive. Guaranteeing connectivity and mobility are some basic requirements for the Internet to evolve.

There are different types of new available broadband networks. The most outstanding are the 4G LTE/Wimax wireless networks that require, as mentioned above, the use of the radio spectrum which is a limited resource. However, deployment of broadband network architectures using fiber optic cables (for instance, Fiber To The Home, FTTH) offer high speed and good performance. Other technology solutions such as the DOCSIS 3.1 cable modem, copper-based xDSL and even PLC power lines provide high-speed access. The technology now available enables global coverage to be provided. Examples include satellite networks, which have currently improved their speed, or undersea fiber optic cables. A variety of technology mixes are available and offer alternative scenarios at the same time. Mobile systems are competing with and replacing fixed networks, which is boosting development of the latter. The quality of the service, either via fixed or mobile connections, is not expected to be an issue in the future. The most suitable technology solution will depend on the characteristics of the demand; based on the services that will be provided, the orography, population density, cost, future maintenance, etc. The optimal deployment choice in each case is a compromise between different technical, socio-economic and political factors.

Backhaul services link mobile services with switching centres within the core network, and the rest of the world, i.e., the public switched telephone network (PSTN) or the Internet. In mobile networks the majority of backhaul services have cell towers connected via fiber optics. In countries that have these ultra-fast connections via optical fiber cables (i.e. in the U.S.), agile deployment of new technologies such as 4G LTE is easier (Federal Communications Commission, 2014a). The need for mobile broadband backhaul support will continue to increase as operators deploy the latest technologies for 4G access that enables users to make intensive use of data via their mobile devices connected to the Internet. Some widely demanded services, such as video streaming, require high-speed connections. In some cases, projections for future growth have been very high, as have been the costs involved in carrying out these investments, which raises new challenges for digital ecosystem firms.

2.2 Emerging Technologies and Future Platforms

Numerous applications and developments appear in the digital ecosystem on a daily basis. This is an environment in which certain actors, which act as gatekeepers, determine the evolution of the ecosystem's dynamics to a great extent. The EED approach helps explain systemic evolution, although changes in the digital ecosystem are often fast and disruptive. The new developments occur at different levels within the Internet: in service infrastructures, integrated platforms, apps, software development and new devices.

Operating systems are one of the factors having the greatest impact on development of new applications. Companies that provide enabling services and solutions to access the Internet, play a key role as gatekeepers in the network. Some years ago it was PCs whereas today smartphones are users' favourite connection interface, which means that a smartphone's operating system is one of the factors that most affects firstly, development, and later access and use of network services based on mobile applications. In the future, the Internet of Things (IoT) will introduce new devices. Two large platforms are outstanding in the digital ecosystem, which are partners with two of the largest business groups nowadays. On the one hand, Apple's operating system with its iOS, and Google (Alphabet) with the Android system.

Thanks to the development of the fourth-generation technology, television broadcasting over the Internet with high-quality video online is changing the multimedia industry, which is facing new challenges and opening new experiences to the users. The arrival of new agents in the ecosystem is changing the audio–visual industry, TV and cinema, similar to what occurred some years ago in the music industry with new business formulas and viewer interaction. The dynamics of these new groups can be analysed under the EED model.

Cloud services (i.e. cloud, TV, etc.) are undergoing rapid growth in the digital ecosystem. Cloud providers have been called the way of the future in network development and access to content such as storage platforms and interconnection of large amounts of content, particularly multimedia. Interconnection and operability between the different cloud services platforms and storage are essential to guaranteeing the plurality of service providers, acting to stop future oligopolies in this field.

The emergence of Big Data enables treatment and analysis of huge amounts of data accessible on the web, resulting from users' interaction on social networks, use of applications, sensors, smart city deployment, traffic mobility patterns, access to public services, energy consumption, distribution networks, smart grids, etc. Available analytical tools enable us to make use of information in very different fields: professional, commercial, operations, intelligence, etc.

The IoT is a new arrival on the digital ecosystem scene that shows the most potential for the near future, particularly for business groups that already form part of the ecosystem.[1] The IoT also shows good prospects for new firms in the ecosystem that want to penetrate a new market with their solutions. The connection of multiple devices used daily or regularly, such as clothes, accessories, health monitoring systems, vehicles, etc. will mean an enormous increase in the number of interconnections and will most likely accelerate infrastructure development. Application of solutions in the healthcare field is transforming medicine, where operation and interaction models with users are evolving.

M2M networks to connect small devices and other equipment are undergoing considerable growth in developed nations. This increase is expected to continue in other areas of business, such as home security, care and other financial, education and leisure services, etc.

---

[1]These companies only need to add functions to their DNA under the EED model, and a growth opportunity through external knowledge to their genomic instructions.

Interconnection and operability between different platforms is a key factor in the digital ecosystem's development. Adoption of standards, homologation of devices and applications have fostered growth and development of the digital ecosystem and have made possible this new environment. This new diversity faces pressures imposed on the ecosystem by big firms, with their pioneering innovative developments, which are ultimately proprietary, and create even greater customer loyalty, as their users unawaringly give their consent.

Security in an open environment like the Internet has become an increasing concern to system users and agents. Progressive encryption of communication via the networks is one of the most accessible solutions. Some cases in point are the development and use of proprietary protocols such as the combination of the protocol developed by Google in 2012, SPDY, with the use of Google's own CPD proxy server. With the use of encrypted data streams and the new protocols, technological advances improve existing security solutions on the Internet. The new procedures make it possible to verify user authentication but also present new problems for global and local security authorities such as government access to communication encrypted by private firms, which have the encryption algorithms, codes and passwords. Technology is aseptic although its use is not. The drive to achieve approval and adoption of the HTTP 2.0 protocol as an open shared standard for all the agents in the Internet value chain can be seen as the response of some system agents to Google's initiative with SPDY. Responses in such a rapidly evolving competitive environment should be even faster or else companies will be left out.

Technology makes it possible to intercept communication by following users' movements from the different communication interfaces to the network. An intrinsic part of the ecosystem is that users must give their formal consent and accept the companies' privacy policy for the servicers used, without feeling they are being controlled by any business organisation.

2.3 End of an Era: An Open Net of Networks?

The Internet model as a sole, open and horizontal network has been evolving since the big business groups achieved control of large enough market shares to secure customer loyalty and capture the users that chose their platform. The development of APIs, open interfaces and standards, or free software are technical features that prevent an abrupt break with the original model of the Internet. Instead, we are witnessing a smoother evolution of the digital ecosystem toward what is practically already a reality in the digital environment. These could be considered small and large fragmented islands in the digital ecosystem. The main ones, which are identified by the operating systems they are based on, still maintain the interconnection between platforms.

These interconnections could break during the ecosystem's evolution. For reasons of security, exclusivity or special features, the networks' isolation and opacity may be accelerated, meaning the de facto break from the model of theoretical free-dom that the Internet represented in its origins.

Rules and regulations applied by governments in their scope of jurisdiction (Open Internet Rules) normally refer to both fixed and mobile wideband access, requiring services providers to publicly and reliably disclose the necessary infor-mation on the management procedures, results and commercial terms or practices they have in place so that users can make informed decisions concerning the use of these services. Keeping watch on an open Internet means adopting regulations that restrict procedures which companies might carry out to discriminate or block access to other agents in the digital ecosystem. The U.S. Federal Communications Commission took a position on this point in 2014 (vid. Federal Communications Commission 2014b).

3 The Value Chain and Business Models in the Digital Ecosystem

The digital ecosystem is a complex environment where agents with extremely dif-ferent activities and roles interact. The Internet's arrival and development have led to new ways of establishing economic relationships on markets. The EED approach can be used to explain the dynamics and adaptation to the environment that the main agents participating and creating the ecosystem have undergone (Gómez-Uranga et al. 2014).

The digital ecosystem is distinctive for being a highly innovative and creative environment where changes are intense and evolution is very fast. If there is one aspect that accurately defines the scenario, it is constant change which is often dis-ruptive and original. Disruptions in this scope enable continuous innovation and technology development.

The new environment that the digital ecosystem is evolving toward involves changes in conventional business models in many industries, but also changes in the roles that were traditionally assigned to market suppliers and users. Roles are sometimes transformed in the digital ecosystem, and this is especially obvious in the scope of content. This is one of the key aspects of the digital ecosystem, not insofar as the volume of business generated at the present time, but due to the high growth and large returns on investments it is showing. Information control is a powerful tool that the industry has been aware of since the Internet's origins. From initial market research to satisfaction surveys, marketing tools have evolved to obtain knowledge of the target users' profiles and even detailed information about them.

Hereafter, we review the Internet's classic value chain model, which was defined by Kearney (2010) at the time. In essence, the model is still valid to describe the main agents that participate in the system (vid. Fig. 1). However, the scope was more limited than the description of the ecosystem, as the participating

Fig. 1 The digital ecosystem's value chain. *Source* Own elaboration based on Kearney (2010)

firms have evolved and continue doing so due to the dynamics the digital eco-system itself is subjected to. Some of the most widely known large multinational firms that have appeared around the Internet, Apple, Google, Facebook, Amazon or Microsoft, can be identified in various links of the value chain. The main agents by type of activity in the ecosystem could be classified in at least one of the following groups:

- *Content owners.* Creators and/or owners of the rights on content which is accessible on the web. They show a mixed profile, ranging from firms that make content generation their business model to individuals that share their experiences, work and creations on the network. The latter do not pursue only economic gain, as they may not receive financial rewards but may achieve social recognition as well as other objectives related to politics, morals, power, popularity, etc.
- *Online services providers.* These companies make all types of applications available to network users, enabling them use IP-based technology communication services over the network such as Voice (VoIP), email, messaging, data, videos. These services raise a challenge for conventional telecommunications operators, due to the technical improvements that have increased the quality of these services over the Internet. They enter into competition with their traditional businesses based on providing landlines and mobile telephony. However, there are also applications that support access to content, such as portals for news, public administration and procedures, leisure, browsers, games, music, shopping, different professional services such as financial, insurance, sales, and purchase platforms, etc. The agents acting within the Internet value chain in this segment usually focus their business model on advertising linked to the use of applications although there are also mixed models that combine free services with advertising and/or payment to access certain services.[2]
- *Technology-based companies and enabling services.* They provide technology support services and applications on the network, hosting web pages or content managers, for instance. They include billing and payment platforms, or advertising enabling platforms such as online agencies or service providers to third parties.

---

[2]Some of them are offered by companies with entertainment business models supplying video, music, game, etc. downloads. Search engines, which is the case of Google, were its main activity in the company's origins (i.e. its genomic instructions).

- *Telecommunications operators and network connectivity services providers.* They offer essential support to establish communication, the core network, providing traffic exchange services or high-capacity information transport services (i.e. highways), backhaul services, access to telecommunications infrastructures for both fixed and mobile networks.
- *User-network interface developers.* They enable contact between users and the network via physical devices such as smartphones, PCs tablets, wearables, as well as applications, software, etc.
- *Providers that offer the technical means and support to develop, deploy and maintain infrastructures.* They are manufacturers of equipment and materials suppliers, test and measurement, certification and homologation laboratories.

The highest growth in the digital ecosystem in the last decade (2005–2015), has taken place in the content segment. It showed the highest figures, registering rates of over 20 %, but lower investment returns than at other links of the chain. This is occurring in spite of the fact that it still only accounts for a small percentage of the ecosystem's total revenues in absolute terms. The next most profitable links in the value chain are online services, technology-enabling services, connectivity, user interface, and service providers with figures under 10 %. The most profitable in terms of returns on investments have been online services and user interface activities, with rates over 20 % in the last decade, followed by content, services providers, technology and firms enabling services and connectivity. Although the latter account for a significant percentage of the total revenues generated in the ecosystem, their business has been seriously damaged and the model is being transformed.

The United States leads growth in the ICT industry, as some of the biggest companies such as Apple, Google or Amazon are North American. Their position in the ecosystem has been strengthened in recent years and their leadership reinforced as a large part of consumers across the world have selected their products. The growing demand for devices such as smartphones and tablets has positioned two companies, the U.S. company Apple and the Korean Samsung as world leaders, followed by the Chinese Huawei and the Korean LG. Meanwhile, past European market leaders such as Nokia (which has now formed an alliance with Microsoft) have gradually lost their positions to the above firms. The case of other European companies that were industry leaders some decades ago was even more dramatic, as they actually disappeared from the ecosystem.

The connectivity segment is one of the most seriously affected by the changes taking place. It is also the most highly regulated due to the public nature of many communications services and the need to use public goods and resources from its beginnings. These agents' business model has clashed with that of the new actors. The unresolved dispute of net neutrality, in addition to others, is the subject of ongoing arguments between agents, with government and oversight agencies sometimes acting as judges and arbitrators as a last resort in conflicts created in the ecosystem.

3.1 Market Transformation. The "Information-Connection" Binomial

The Internet is a powerful tool for communication, opinions, information and interaction between users across the world. The digital ecosystem's development has and will continue to be a key driver in the overall process of globalisation. Globalisation of the economy is a complex process of interdependence between the different countries, boosted by increasingly easier communication and interaction, and international agreements to eradicate barriers to world trade. Commercial, financial transactions or exchange of any product or services can be carried out in the digital ecosystem because there are no borders. As production delocation is undergoing in terms of people, capital and technology, and all of these are also connected to the network, they too are undergoing delocalization.

The Internet has changed the way we understand commercial relationships. User connection capacity with increasingly user-friendly high-performance devices at reasonable prices available to a large sector of society has played a key role in this new model. Regardless of their role in the ecosystem, they can all access the web. There are no entry barriers other than those allowed by the infrastructures and connection devices in this environment free of conflict with the authorities. Governments can block or limit connection to the network from a particular physical location in a country, where the government controls connectivity or even imposes sanctions on users for certain activities in the digital ecosystem. However, they cannot eradicate the ecosystem.

The Internet, which emerged as a tool to solve a problem in the academic field, has become a complex all-encompassing system. Practically, all types of activities can be found on the network. Internet covers everything from the professional to personal scope of each user that connects, regardless of age and origin. The network is multicultural and diversity is a key part of the Internet, which offers content and services for all types of public. This global connection capacity, linked to the possibility of interaction in real time between network participants, is a disruptive combination if we take into account that its scope is also global. The only borders on the network are language and culture.

The Internet's peculiarities have solved some market failures while also creating new problems and challenges. Many of the information asymmetries existing on markets have only been partially solved by the digital ecosystem, where information appears on many webs, forums, recommendation services, etc. Interconnection platforms and online shops enable contact between suppliers and users requiring all types of solutions. However, information is not knowledge. Agents' roles may also be ambiguous; i.e. the same user may be a content generator that uploads to the network, while also demanding similar services from others.

The network's availability and ubiquity provides a new scenario where there are no timetables or barriers to business. Permanent interconnection is possible on the net, thus modifying conventional market rules. Its global nature clashes with each country or state's traditional model, which has specific legal systems (such as tax regulations, intellectual property). This model has altered the different countries' taxation schemes on transactions. As a result, many economic sectors have benefited. Transport is a case in point. It has significantly improved with new telecommunications technology to manage logistics and transport fleet traffic, and in turn, fostering and promoting e-commerce.

The most relevant platforms at the present in the Internet ecosystem are mainly led by American and some Asian firms. Other developed regions such as Europe and the most advanced Asian nations, Japan for example, have not managed to capture a significant part of the value that the ecosystem has been creating with new business possibilities in information technologies, especially among services and applications providers.

Existing laws in each of the countries, at the national or regional level, such as the European Union, and their application to the new markets that have emerged and developed on the Internet have played a key role in spreading the success of some of today's leading global firms (see Chapter "4G Technology: The Role of Telecom Carriers"). This may also explain why the conditions in the environment have not fostered new alternative rival firms in the industry to compete with the leaders. The legal and regulatory differences, together with other cultural factors such as innovation support are important conditions in the environment when explaining the transformation of markets and their agents, as well as the evolution of the ecosystem. Easy access to capital markets to obtain the necessary funding, or even the intellectual property protection system, contribute to understanding some of the keys to the top firms' transformation and success in adapting to the environment.

3.2 New Applications and Services

From the economic point of view, the content industry is the key future segment in the digital ecosystem. This industry will have to undergo evolution until it finds the most suitable models to enable content creators to correctly position themselves in the ecosystem. In the current environment, it is the gatekeepers that occupy a privileged position, constantly exercising their power to foster or restrict access to content via online services and different interfaces and their own platforms. They act as intermediaries on a two-sided market between users and creators or owners of the content rights.

As mentioned, mobile applications are the new Internet ecosystem services showing the fastest growth. Mobile wideband traffic is reaching levels that are displacing fixed wideband connections. These services create revenues for applications developers and also from the sales generated on the network, in addition to the advertising they include or applications sold on the web. Online services providers are registering very high growth in comparison to the rest of the links in the Internet value chain, except for the content industry. Its main companies were showing higher than double-figure growth in 2014 (between 15–30 %).

This is also occurring with the returns on assets, which exceed 10 % in many companies, far above the connectivity-related activities in the value chain. Telecommunications operators usually have very low return rates, under 10 % in the most favourable cases (around 2–4 %) in 2014.

Estimates made by Vision Mobile and Developer Economy indicated that the activities created by the mobile applications market generated a total value of $60–70 billion on the global market. According to estimates made by the Boston Consulting Group (2014), in only five European countries, (Germany, France, Italy, Spain and the United Kingdom), the total revenues from Internet mobile applications, content and associated services reached €33.12 billion in 2013, which is 36 % of the total revenues of €92 billion. This figure takes into account the activities related to the rest of the links in the chain (access offered by mobile services operators and providers, revenues generated in enabling platforms devices and operating systems for mobiles and the operating costs related to maintenance and expansion of the network and infrastructures). The Boston Consulting Group estimated total revenues of €226 billion in 2017 for all the activities generated in the mobile ecosystem in the European countries mentioned (ibid). This means growth of 25 % in 2013–2017, considering a smartphone penetration rate of 64 % for adult users.

An enormous variety of mobile applications are available, from search engines, a basic tool and one of the most widely demanded in 2014–2015 by smartphone users that surfed the net with their device for music, video, leisure, games, trips, localisers, news portals, etc. Email, communications services such as WhatsApp or WeChat, and the applications classified as social networks (e.g. Facebook, Twitter, Instagram) were the most popular on the web in 2014. Many analysts agree that education, where there are already many applications, and the healthcare industry, which is expected to undergo sharp growth and development of new applications, are promising. The financial industry is also registering changes in the ecosystem, with a greater impact on an increasing number of national economies. Mobile applications to access banking services are widely used in many countries. These services evolve rapidly and offer more complete options, which is also the case of mobile payment systems with platforms and different devices or wearables. Many users shop on the web via their mobile devices (i.e. smartphones or tablets).

Most of the applications designed for mobiles can be easily downloaded via web browsers, from platforms or virtual shops for mobile applications such as Apple's App Store or Google Play, both of which are very popular. The number of available applications for Android operating system users alone was over 1.3 million apps in 2013. The Apple platform ranked second worldwide with 1.2 million (Federal Communications Commission 2014a). High-speed connectivity to the net may not be required for their use, depending on the application. However, they may also be operative without the need to connect once they have been installed in the user's terminal. It is increasingly common to find pre-installed applications in the operating systems when purchasing a terminal.

Another new development that appeared in the digital ecosystem is virtual currency (i.e. the bitcoin). It could mean a disruption in the macro economy and the way we understand monetary policy. Electronic trading and currency and securities market platforms have been operating for years in the digital ecosystem. The first Electronic Communication Network (ECN) was created by the American National Association of Securities Dealers (NASDAQ) in 1971. The leading ECNs now trade billions of dollars in shares and currencies on a daily basis.

The applications related to multimedia content and social networks are a powerful combination. The new Cloud TV services with gatekeepers that can pre-empt the market and displace the TV industry to the network with the possibility of horizontal and vertical integrations with other agents, raise new challenges (vid. Noam 2014; Waterman et al. 2013). Multimedia applications for live video streaming such as Periscope or Meerkat are one of the latest tendencies in apps technology. They follow in the wake of other highly successful apps such as Instagram or Twitter where users share opinions and photographs.

The IoT has raised expectations concerning the scope of integration that new devices linked to everyday objects and items can reach, in addition to the connectivity and interaction challenges between devices. M2M communications have increased considerably in recent years, providing connectivity to different devices, sensors, etc. which are connected to the network and usually found in professional environments. However, M2M is showing a tendency to go beyond professional spheres and move into homes.

3.3 Relationships Between Agents. Interaction and Role Transformation

The evolution of business models, cash flows and user empowerment are transforming the way in which participants are organised in the environment. Globalisation is one of the key factors in analysis of the digital ecosystem. Disruptions are rapidly transferred across the world. The impact of changes and innovations in any activity is displaced throughout the ecosystem with results that are sometimes surprising. The scope and far-reaching influence of certain business groups in this environment give them an especially relevant role, not only in their primary field of activity but in the entire ecosystem. This, in turn, enables them to further strengthen their power if their long-term strategy is successful. The EED model makes it possible to analyse the case of some of these companies such as Google or Apple.

The so-called gatekeepers play an essential role in keeping the digital ecosystem running properly. The dynamics of the rest of the agents in the system and their evolution may be analysed under an epigenetic perspective. Companies that act as gatekeepers have become vital. Their actions filter access to other ecosystem activities, determining and even preventing the participation of other agents that are unable to overcome the access barriers if users choose said gatekeepers' applications and platforms.

The role of gatekeeper can be filled by large business groups that operate in the digital ecosystem. Google is, for instance, the case of a perfect platform. The fact that only a few platforms are competing for the lion's share is a factor that increases their power. The critical size reached by these firms, which is combined with the multiplier effect of network-based applications in which Metcalfe's Law can be applied, ultimately condition the dynamics generated in the digital ecosystem and its evolution. Multinationals such as Google, Amazon and Apple have become global agents with privileged positions in the ecosystem in just a few years. This arouses suspicion and creates alarm in the environment where other agents feel overwhelmed and threatened by the transformation in the ecosystem (see Chapter "The Digital Ecosystem: An "inherit" Disruption for Developers?" by Vega et al.).

Google is the world's leading web search engine. Its privileged position has placed certain regions on alert, such as the EU where Google searches accounted for 92.5 % of all search traffic at the beginning of 2015 (a market share which is even higher than its original market, the U.S.), in comparison to other competitors such as Bing 2.6 %, Yahoo 2.2 % or Yandex 1.3 %, while the rest of the platforms controlled 1.6 % of the market. The scenario has prompted EU institutions in charge of overseeing competition to accuse the company of abuse of market power in its web search business. However, it is not the only battle that the American giant is involved in and could be one of the reasons behind Google's new organisation structure and new name: Alphabet. Like other big leading Internet ecosystem companies, it has diversified considerably in recent years, from search and ad business to self-driving cars and life sciences research (see Chapter "Epigenetic Economics Dynamics in the Internet Ecosystem" by Zabala-Iturriagagoitia et al.).

An unfavourable political climate with legislative changes in many countries, above all in Europe, concerning data protection, guarantees for intellectual property and patents has led to disputes, many of which are pending resolution. The rest of the agents affected in these disputes, including the end users of applications, sometimes have conflicting views due to the very synergies and collaboration which are established between business groups in the ecosystem. It is remarkable that users' views are often closer to private companies' positions than their democratically elected governments'.

In 2015, Amazon was the world's largest e-commerce company. As per cloud services, Amazon led with nearly 28 % of this market at the end of 2014, which was only a decade after it launched its cloud computing business service. It is followed by other Internet giants such as Microsoft with 10 % at the end of 2014; IBM, Google and Salesforce, according to information provided by Synergy Research Group. Cloud services are growing very quickly, yielding over double-figure returns in some cases. In 2014, this return was 51 % higher for Amazon than the previous year.

From the legislative and legal perspective, there is no global judicial framework that establishes common rules or a shared law. In practice, this avoids carrying out systematic preventive control of the network. Legal and political systems vary according to the country and we may find very different decisions depending on the country where judgement is handed down. Certain activities persecuted by law in some countries are not in others. Agents such as Google or Apple are expanding toward new businesses in the global ecosystem, which, in some cases, are under the authority of the communications services that the operators provide in each country. This has caused alarm, not only between these giants with feet of clay but also in government security agencies which have been allied with operators and companies responsible for providing network connectivity (see Chapter "4G Technology: The Role of Telecom Carriers").

Technology developments such as new protocols are sometimes released by those firms. Some may improve user navigation or automatically encrypt the user information transferred when they access certain nets and platforms or afford the possibility of routing heavy traffic flows to private proxies. All this raises new challenges for intelligence and government services that will be forced to negotiate on a scenario where big companies that have the control may not be under the jurisdiction of the country where the conflict originates.

4 Internet Governance

On a complex open scenario like the Internet, where multiple agents are constantly participating and interacting, the issue of international governance calls for large amounts of creativity and innovation. Internet is a space open to diversity and its governance should follow suit, based on inclusive participation, preserving, boosting and developing cultural diversity. The model of an open, distributed, self-run Internet has led to a working system which is unusual in other scopes. It is distinctive for multi-stakeholder participation as opposed to the multilaterality-based mechanisms that have been in place for decades in telecommunications, such as the case of the International Telecommunications Union (ITU).

Multilateral cooperation principles are based on the participation of all the agents (government, the private sector, civil society, and the technical community) in equal conditions. Participation is fostered and procedures that guarantee equal opportunities to collaborate and contribute to the decision-making processes are adopted. Transparency is a basic rule in all the processes, which are duly made known, public and open. Responsibility is another key point and involves implementation of accountability mechanisms to verify the results obtained. The approach is always based on consensus and must reflect the different points of view held by the entire community. This is an extremely ambitious cooperation mechanism but proves difficult to translate to reality which is riddled with interests, double talk, disputes over representativeness and legitimacy in decision-making. In actual practice, all of the above mean that this model does not function effectively, although it is an interesting experiment which, nevertheless, could solve many of the conflicts found in the ecosystem.

The support that many governments have provided to develop the current Internet cannot be denied. From the origins of the Internet, the U.S. government has played a crucial role, enabling its creation and evolution to the present moment of transition. The role of governments is pushed into the background in the multi-stakeholder cooperation model as it facilitates deployment of telecommunications infrastructures that make it possible to create an optimal environment for access to the ecosystem. It envisages regulatory frameworks to achieve this but avoiding a proactive protectionist role that could limit innovation. Another of the tasks that falls under governments' responsibility is developing a legislative framework in their respective countries that strengthens the idea that the Internet is not an area outside the law where anything is allowed. This requires that the legislative framework evolve at the same pace as the ecosystem, which, as experience has shown in the last decades, is difficult to achieve. The conflicts that arise on the network often do not have the proper legal framework needed to resolve them in a systematised manner.

Human rights, freedom of expression and dissemination of information on the Internet or protection from cyber terrorism are other extremely important topics where governments must play a role and which have been debated in international Internet governance forums in recent years. Note that all these dimensions are included as the analysis of the consequences (in terms of innovation) as a result of epigenetic factors in the EED approach (see Chapter "Introducing an Epigenetic Approach for the Study of Internet Industry Groups").

Cooperation between governments, industry and society may lead to public policies that address the true needs of the digital ecosystem more directly. There is no clear group of leading agents in a multi-stakeholder environment. Governments, which are accustomed to being the leader, face a new framework, Internet governance, in which they participate on equal footing with the rest of the agents. This is a very different scenario from their participation in multilateral forums. Many find this contradictory, as they consider themselves the legitimate representatives of the citizens who elected them, at least in democratic nations.

It is hoped that governments will support the multi-stakeholder governance model and pass national laws that include the measures needed to safeguard the Internet and its freedoms, in search of the complex balance between the different parties' interests, which are often difficult to reconcile. The Internet Governance Forum (IGF) is the main international forum, and was created in the spirit of a multi-stakeholder approach, fostering internationalisation of Internet governance and cooperation. The first official IGF event was held in Athens in 2006. Since then, yearly meetings are held in which thousands of people representing all the stakeholders debate Internet governance. Another more recent forum is NetMundial, first held in Brazil in 2014. At this event, the community defined a series of Internet governance principles and began a new push for governance channelled through the yearly IGF held shortly afterward. One of the essential Internet governance principles, agreed by the different interest groups, looked to strengthen the Internet's basic values, advocating its recognition as a global resource that should be managed in the public interest. The focus was on achieving regulation for the Internet that would guarantee citizens the same rights in the online world as in the real world, in agreement with the Universal Declaration of Human Rights and the human rights obligations established in international legislation. The rights recognised included freedom of expression, association, privacy, accessibility, information and access to said information, to development of countries by promoting innovation, creativity, innovation and cooperation between stakeholders.

4.1 Regulation of the Digital Ecosystem

Regulation of the net is an extremely complex task that goes beyond the virtual borders of each country's legal framework. Guaranteeing security and continuity of the original network design requires international cooperation between all countries. In future, the digital ecosystem should preserve and guarantee the original principles of freedom, openness and neutrality.

The convergence of conventional telecommunications with the computer industry created a fruitful combination that triggered the creation and later development of the digital ecosystem. Both industries had very different starting points. The traditional telecommunications industry has been regulated under the control of each country's telecommunications operators from its very origins. In many cases, they were directly supervised by government. However, the computer industry's origins had an open, global, more competitive and innovative approach and were not subject to public service obligations, use of public resources and goods and costly infrastructure deployment. Both industries have competed in the digital ecosystem although a new optimal regulatory and legislative framework to manage the new ecosystem has not been found. This has created regulatory asymmetry which has become even more obvious as the system has expanded and new business models and groups have appeared. Nevertheless, regulation of communications infrastructures has continued to evolve and has become, in some cases, excessively complex and unsuitable for this particular stage of the network's evolution.

The debate on regulatory asymmetries in the digital ecosystem and their solution have been demanded by operators for some time now. Other debates related to competing business models have arisen around this unsolved conflict. One such debate is network neutrality and the regulation of big Internet groups' businesses, for instance Google or Apple. Many telecommunications operators are incumbent in one country or several, but not global, as opposed to firms in other businesses which create value on the Internet. They are calling for a regulatory framework similar to that of their new competitors, the online services and enabling providers, applications and services platforms. Ultimately, they are demanding non-intrusive regulation or even better, one that allows agents to regulate themselves.

4.2 Network Security and Control

In non-democratic systems, governments are often tempted to control the Internet, taking "information is power" as their motto. However, they end up clashing with the network design. Censorship on the Internet is difficult to eradicate; some governments distrust citizens' freedom of expression and their self-organisation abilities. This could pose a threat to existing power models. It is possible to force disconnection from the network but not to eradicate the Internet. It is also feasible to build a new network with a new made-to-order ecosystem, which is specific to a certain country or geographic region, similar to many of the islands that exist in professional fields on the net. However, it would be very different. Interconnection with the global network is what users are demanding and is what makes the Internet ecosystem so unique.

As Castells (2003) pointed out, as the use of the Internet has spread, information and social behaviour toward the Internet have become more important. Control over the Internet, the battle for freedom on the network has shifted from being the exclusive concern of the old elite and has become widespread. Access to the network is protected and guaranteed in democratic countries. This is based on freedom of expression, citizens' rights over public resources and goods, etc. as established in countries' constitutions and legislative and judicial systems. The debate on whether access to the Internet should be considered a universal human right, or the Internet as a global public good, raises divergent opinions on the scope of the digital ecosystem. Some people consider the Internet a technological tool to exercise fundamental rights. In this sense, the European Union pointed out at the start of 2015 that the use of the Internet was an inalienable civil right whose application national authorities could contribute to within their competences frameworks (Vid. Official Diary of the European Union 21.1.2015).

There is deep contradiction between the freedom provided by the digital ecosystem and the control and vigilance on citizens' lives that technology also allows. Users' propensity to lose control of the network in benefit of private firms or security agencies and governments elected in democratic processes is a tendency that cannot be reversed. Users lose their privacy on the network on a daily voluntary basis each time they accept the security and data protection disclaimers required by most of the applications they use. Connection and constant interaction on the network, combined with the possibility of being constantly geolocated on mobile networks offer many advantages and are creating new interaction models and added services. However, this carries high costs as regards privacy which, however, many users are not capable of assessing proportionately.

In spite of the different countries' approaches to the digital ecosystem, the economic and geostrategic role of this new ecosystem is so vast that, in actual practice, none can fail to enter and take part in its evolution. The rift existing between countries concerning the global Internet governance model was made evident in 2012 when numerous countries decided not to ratify the regulations on international telecommunications, which was put to the vote following the World International Telecommunications Conference organised by ITU in Dubai in 2012. Later events followed, such as U.S. Intelligence agent E. Snowden's disclosure in 2013 of some of the ecosystem's weak points concerning privacy, security and abuse of power on the Internet by some governments, particularly the U.S.

The events prompted response from many emerging nations and other countries such as Russia and China which had not been aligned with the U.S. government's position for some time and questioned the role of the U.S. as the safe haven guaranteeing freedom of expression on the Internet. They also criticised the Internet technical model of governance, and the critical resources control, which had until that time been under the responsibility of a group of U.S.-based private institutions supervised by the U.S. Department of Commerce.

Critical Internet resources include technical standards on protocols, procedures and services, the infrastructures needed for the net to work, root servers, assignment of domain names, addresses, IP or protocol administration. All of these resources are managed from private organisations: Internet Corporation for Assigned Names and Numbers (ICANN), Internet Assigned Numbers Authority (IANA) and the Root Server System Advisory Committee (RSSAC), (these two latter bodies are under the ICANN, although they operate autonomously), VeriSign, the Internet Engineering Task Force (IETF) and the World Wide Web Consortium (W3C).

Root servers are located across the world; only 3 of the 13 active servers in 2015 were outside the U.S. The primary root server that keeps a copy of all the Top Level Domains file of the other 12, is also based in the U.S. Management of these resources which are critical to the ecosystem's operation is decentralised and the different stakeholders are represented in the ICANN governing body. Although this model has functioned to date, it is inconvenient for many governments and agents, which are calling for greater independence and decentralisation. Governments are represented in ICANN through the Government Advisory Committee (GAC). This participation has always been considered limited and in recent years, a greater imbalance of powers between countries has been perceived.

Against such a background, reorganisation of the control of critical Internet resources became a topic of debate in 2014. Revision of the assignment of top-level domains was accepted by the U.S. government, shifting from the unilateral assignation model that had been used through ICANN and IANA until that time to a new collaboration model. The U.S. accepted the criticism expressed by numerous stakeholders and some allies at different forums and international meetings on Internet governance.

These conflicts and other new ones caused within the ecosystem entail significant changes in the environment and may put an end to the model that we know, obliging companies to adapt as shown in the EED model and taking the ecosystem to a new stage. The risk of fragmenting the ecosystem into different blocks can only be avoided by promoting transparency in Internet management and policy. This requires technologically solid architecture and infrastructures that give confidence to users and security to governments, protecting and improving the resilience and the security of the Internet. The virtues of the Internet ecosystem can be preserved and further advancement can be achieved through respect and promotion of cultural diversity and the safeguarding of human rights.

References

Castells, M. (2003). Internet, libertad y sociedad: una perspectiva analítica. *Polis, 4*, 1–17.

Cerf, V. G., & Kahn, R. E. (1974). A protocol for packet network interconnection. *IEEE Transactions on Communications, 22*, 639–648.

Cisco (2015). Cisco virtual networking index: Global mobile data traffic forecast update, 2014–2019.

Clarke, R. N. (2014). Expanding mobile wireless capacity: The challenges presented by technology and economics. *Telecommunications Policy, 38*, 693–708.

Federal Communications Commission (2014a). 17 Annual report and analysis of competitive market conditions with respect to mobile wireless, including commercial mobile services. Washington.

Federal Communications Commission (2014b). Protecting and promoting the open internet. GN Docket 14–28, Notice of Proposed Rulemaking, 29 FCC Rcd 5561.

Fransman, M. (2010). *The new ICT ecosystem*. Cambridge: Cambridge University Press.

Gómez-Uranga, M., Miguel, J. C., & Zabala-Iturriagagoitia, J. M. (2014). Epigenetic economic dynamics: The evolution of big internet business ecosystems, evidence for patents. *Technovation, 34*(3), 177–189.

ITU (2015). ITU World Telecommunication-ICT Indicators database.

Kearney, A. T. (2010). Internet value chain economics: gaining a deeper understanding of the Internet economy. Report for Vodafone.

Kuhn, T. S. (1962). *The structure of scientific revolutions*. Chicago: The University of Chicago Press.

Leiner, B. M., Cerf, V. G., Clark, D. D., Kahn, R. E., Kleinrock, L., Lynch, D. C., et al. (1997). *Brief history of the internet*. Internet Society.

Leiner, B. M., Cole, R., Postel, J., & Mills, D. (1985). The DARPA internet protocol suite. *IEEE Communications Magazine, 23*, 29–34.

Noam, E. (2014). Cloud TV: Toward the next generation of network policy debates. *Telecommunications Policy, 38*, 684–692.

The Boston Consulting Group (2014). *The Mobile Internet Economy in Europe*. BCG.

Waterman, D., Sherman, R., & Wook Ji, S. (2013). The economics of online television: Industry development, aggregation and TV everywhere. *Telecommunications Policy, 37*, 725–736.

4G Technology: The Role of Telecom Carriers

Andrés Araujo and Itziar Urizar

1 The Mobile Phone Sector

Mobile technology has been performing quite dynamically since 2008. User levels are on the rise around the world, as are connections and information traffic. In the 4 years from the end of 2008–2012, the global mobile market grew at a rate of 13.7 % p.a., the number of individual subscribers increased by 38 % and the number of Internet connections by 67 % (Page et al. 2013).

Today, smartphones are habitually used to listen to music, download videos, send mail, access social networks or to shop, to the extent that they are now an essential access point to new business models and the gateway to the Internet ecosystem (Fig. 1).

In fact, as the below figure shows, mobile devices (basically smartphones and tablets) are rapidly replacing desktops, laptops, and consoles as a means of access to Internet. In barely 5 years, mobile devices have leapt from less than 5 % of Internet traffic to a share of more than 37 % and, as Fig. 2 shows, Internet data traffic via mobile devices is expected to continue to enjoy major worldwide

A. Araujo (✉)
Department of Management and Business Economics,
University of the Basque Country, UPV/EHU, Bilbao, Spain
e-mail: andres.araujo@ehu.es

I. Urizar
Department of Applied Economics I, University of the Basque Country, UPV/EHU,
Bilbao, Spain
e-mail: itziar.urizar@ehu.es

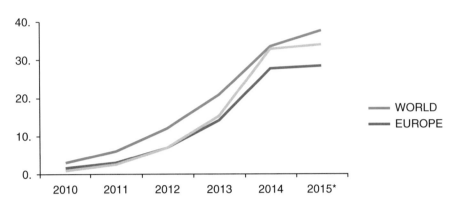

Fig. 1 Share (%) of Internet mobile device traffic. *Source* Own elaboration based on stat counter global stat data. *Note* (*) data up to April 2015

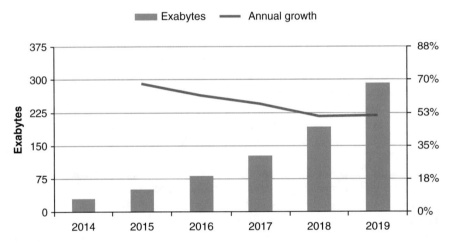

Fig. 2 Forecast of mobile data traffic. *Source* Cisco visual networking index. Gerhardt et al. (2013)

growth. Cisco (Gerhardt et al. 2013) forecasts a compound annual growth rate of 57 % to 2019, when it should reach almost 300 exabytes.[1]

The sector's huge global growth has come about largely in Asia, Latin America, and Africa. Asia generated 57 % of new connections, with Latin America and Africa sharing 20 % (Page et al. 2013). Falling telephony prices, the lack of viable alternatives (fixed lines) particularly in Africa and major investments by carriers, also known as operators, to improve broadband infrastructures are all major factors driving growth.

---

[1]An exabyte (Eb) is equivalent to 1000 Pb (Petabytes) or 1 million terabytes (Tb).

Between 2011 and 2013 average smartphone prices fell 25 % and are expected to drop another 19 % by 2017 (Bock et al. 2015). Performance in terms of cost and data consumption over transmission speeds has also improved significantly.

Investments in mobile technology are still huge; from 2009 investment levels have been higher than the pharmaceutical and biotechnology industries combined. Mobile Internet generates an enormous volume of investment driven by the whole of the ecosystem, operators, software producers, hardware producers and contents, services and applications providers included. Mobile's contribution to GDP puts it up there amongst the leading sectors in the economy, particularly in the United States and China, as Table 1 shows.

Growth is expected essentially to be driven by the contents and app subsector and to generate extraordinary value in terms of consumer surplus. In Europe, in per capita terms in the UK, Germany, Italy, France, and Spain, the difference between what users are prepared to pay and what they really pay is €4700 and represents nearly 13 times more what they pay in these countries (Bock et al. 2014: 18). More than 50 % of this consumer surplus is generated by Android and roughly 30 % by Apple iOS. In general, in these European countries an adult spends an average of €555 on telephones, tablets, data transmission, digital contents, and e-commerce a year.

Another major factor is the employment the sector as a whole generates, basically by creating knowledge-intensive jobs linked to high qualifications. In Europe, in 2013, employment in the mobile app sector was 1.8 and for 2018 it is expected to be 4.8, almost three times more (Fundación Orange 2014).

However, the change to 4G is still on the agenda. This new technological benchmark is revolutionizing the information society by accelerating and enabling increasingly large sectors of the population to access Internet from mobile devices thanks to the next generation of broadband networks (i.e., *Next generation Networks* or NGN) or ultrafast networks.

The prospects for global growth that 4G Long-Term Evolution (4G-LTE) technology brings to the sector are positive, despite penetration being minimal so far and that in Europe there is a regulatory problem concerning the radio spectrum. To date growth in the number of mobile connections has been due largely to 2G and 3G but it is estimated that by 2017 the change from 3G to 4G will mean a

Table 1 Mobile telephone sector's contribution to GDP (2014) and growth rates (2009–2014)	Contribution to GDP (2014) (%)	Growth 2009–2014 (%)
United States	3.2	15.4
Germany	1.8	9.1
South Korea	11	18
Brazil	1.7	11.7
China	3.7	17.7
India	2.2	12.4

Source Bock et al. (2014)

substantial increase. In 2017, one of every five mobile connections will be via 4G as opposed to the 1:25 registered in 2012 (Page et al. 2013). With these prospects, by 2020 mobile broadband will account for 90 % of global data traffic (Ericsson 2015). Such figures bear witness to the huge growth potential in the sector and in other economic sectors, considering the opportunities for innovation associated with education, healthcare, e-commerce, and even public administration.

Nevertheless, given that progress in the mobile technology sector to date has been due more to the development of previous standards like 2G and 3G than to the actual implementation of 4G, we need to understand the adaptive changes now occurring in a tumultuous, uncertain environment like the Internet ecosystem. To that end, we shall be taking an evolutionist, epigenetic look at the performance of operators, service providers, and major Internet groups (see Chapter "Introducing an Epigenetic Approach for the Study of Internet Industry Groups" in this book by Gómez-Uranga et al.). In response to changes in technology and the introduction of 4G, such players have swiftly modified their original business routines, giving rise to new identities and business strategies (Gómez-Uranga et al. 2014).

Prior to that, though, in Sect. 2 we provide a brief technical description of the new technology. In Sects. 3 and 4 we tackle the regulation that has favored or in some cases hindered the development of the telecom sector and the digital economy in Europe in general and in Spain in particular. Section 5 addresses some of the epigenetic forces being unleashed in the ecosystem. Section 6 is central to this particular chapter. In it we explore the role telecom carriers play in the Internet ecosystem, analyzing operator strategies and performance in entering activities such as the entertainment industry, financial services, or e-health. Finally, the chapter concludes with a discussion of future trends and challenges.

2 The New Technology: 4G

The fourth generation mobile communications Long-Term Evolution (LTE) standard, called 4G, is the latest stage in the evolution, from the 1970s, of a series of standards that have been through a number of phases. The first phase, called 1G, was essentially bound up with the development of the mobile phone and voice communications. Throughout the 1990s, the 2G era entailed the development of the Global System for Mobile communications (GSM) standard which also implied data, and not just voice, transmission. 3G arrived between 2000 and 2010 with the development of the Internet. But the need to transmit increasing quantities of data and information in the shortest possible time led to the emergence of 4G or the LTE standard which was finalized in 2007 and first launched by TeliaSonera in 2009 in Scandinavia. From then on, there have been shifts and movements on the market and alliances between the main patent owners (e.g., Qualcomm, Apple, Ericssson, Samsung) to commercialize 4G-related products (Han 2015).

4G basically means lower latency time when smartphones access Internet, a greater flow of data or information and faster data transmission speeds for clients, greater than 100 Mbps with the implications this has for the demand for contents, new applications, and services. Even so, some mention is required here of the major analytical difficulties in the technical description of the new fourth generation standard, in respect of the previous technologies or standards (Curwen and Whalley 2011), particularly from the political economy perspective adopted in the present discussion.

In general terms, the 4G, fourth generation technology is a standard for mobile communications established by the Third Generation Partnership Project (3GPP) as an update of the potential of 3G technology. Improvements over the latter technology basically involve an increase in information flows and transmission speeds, i.e., data flow transmission rates of between 1.5 and 3 Gbps.

Key factors in 4G technology are the frequency band and bandwidth. The lower the frequency (800 MHz), the greater the coverage for a given distance (up to 5 km with antenna) and the greater the quality of the actual transmission. As the frequency increases, losses and errors in signal transmission lead to loss of quality in the communication. Further, the higher the frequency, the lower the penetration capacity. The greater the bandwidth, the greater the flow of data for a given quality of service. The real coverage of each antenna (base station) depends on these two factors. When population density is high, greater bandwidth is required for a given frequency band, while in rural areas what determines the number of base stations is the physical distance achievable.

Siting is a decisive factor in deploying wireless 4G networks and, specifically, for base station installation, when coverage and the quality of communications need to be considered. The degree of urban development in a particular area and the accompanying demographic density are thus key aspects of regulation in promoting the development of 4G technology, as are the number of mobile telephone users and the number of users with smartphones with sufficient capacity.

For example, frequency bands in Spain today are 800 MHz, available as from January 2015. The 1800 MHz band is shared with 3G and 2G users. Although the 2000 MHz band is available it lacks the range required and starting to invest in it from zero may be more expensive for network deployment. The 800 MHz frequency band is suitable for thinly populated areas providing sufficient coverage and range in rural zones and enabling access to broadband services via wireless technologies, while also helping to narrow the digital gap.[2]

With new frequency bands or new spectrum resources also becoming available thanks to the digital dividend, new business and innovation opportunities will be enabled in healthcare and educational services and in others will facilitate electronic access to public administration or the Internet of Things (IoT). Finally, the bands used by the actual mobile and the 4G category itself are factors limiting the quality and deployment of 4G technology. Like the operators, not all mobiles use the same frequencies. For instance, the LTE iPhone 5 does not connect to the 2600 MHz band (band 7) used in some cities where the 1800 MHz are saturated

---

[2]In Spain, Yoigo (TeliaSonera) has been the most active carrier in deploying 4G while Movistar (Telefonica), Vodafone, and Orange have waited for the liberalization of the band before using the digital dividend to drive the expansion of 4G throughout the country.

and would have to wait for the arrival of 800 MHz. As far as the 4G category is concerned, category 3 offers up to 100 Mbps and category 4 up to 150 Mbps.

So when it comes to grasping the greater business potential linked to 4G technology in economic activity as a whole, it is essential to consider, together with the need for better ultrarapid networks, the importance of having the best terminals. In other words, we need to take a detailed look at the group of players that interact in this Internet ecosystem and the epigenetic forces produced with a view to upping competitive levels in the market and economic profitability (see Sect. 5 in this chapter).

Before discussing these new forces and the adaptive changes now occurring in the ecosystem, however, we are going to detail the development process and telecom sector regulation in Europe and Spain.

3 Regulations: Some Issues

From its origins in the 1980s, the mobile technology market in Europe has been very fragmented from the technical point of view, with a huge variety of technical standards. From the business perspective, it has been an oligopolistic game. It was not until the 1990s that we began to see some attempts at liberalizing the sector with the introduction of the GSM digital system to create a single European mobile telephone standard. Thirteen countries agreed to create a pan-European network with the GSM. In Spain, the process met with major delays, as the third GSM operator (formerly Amena, now Vodafone) did not begin business activities until 1999, a year before the new 3G licenses were awarded.

GSM brought SMS and Internet access with it, via the mobile telephone. It facilitated increased geographical coverage and massive use of the service. In 1996, the European Commission assumed leadership of the liberalization process and issued a directive to abolish exclusive rights on the use of infrastructures and to promote interconnection and interoperability while also attempting to increase the number of operators. This new regulation obliged member states to issue at least two GSM licenses and not to limit the number while frequencies were available. It enabled carriers to create their own networks and guaranteed them interconnection with public networks.

In a bid to develop a new mobile telephone system, European countries joined forces and in 1999 launched the new 3G system. The first licenses were issued in 2000. Between 2010 and 2012 there was a glut of tenders put out owing to the growth in demand for mobile telephones and, above all, for data transmission. The first countries to put licenses out to tender (Germany, UK, and Italy) attracted important offers; later ones, like Spain's, got poorer results. While the UK collected some €642 per capita, Spain only collected €11 (Ganuza and Viecens 2014). Subsequently, the crisis in the sector had a drastic effect on investments. Technology that was still immature at the time of license issues, and the lack of terminals, also slowed the spread of the new system in Europe.

Shortly after 3G began to be commercialized, European institutions started work to develop 4G for high-speed mobile communications. Finland was the first European country to assign a spectrum for the development of 4G in 2009 (Curwen and Whalley 2011). In 2009, the Commission approved a new regulatory framework for all electronic communication networks and services (Directive 2009/140/EC from the European Parliament and Council) with the aim of progressing toward the creation of a single telecommunications market. National authority functions were strengthened under the principles of competition and the protection of citizens' interests to achieve greater transparency and efficiency in the allocation of frequencies. The idea was also to improve service quality and ensure contractual obligations to clients were respected. The issue of new licenses, initiatives contributing to better phone number portability, commercial policies encouraging terminal spread amongst users and telephone price reductions were also designed to improve efficiency in the sector and open the market up to competition (Huertas 2013).

Regulatory policy has brought risks associated with uncertainty in the operators' working environment, particularly as the sector is subject to regulation in countries with different legislations. Possible modifications to the conditions for obtaining and renewing licenses and concessions (i.e., delays, tariff increases), the possibility of limitations being established on the maximum earning percentage, of limits being placed on the prices of goods and services, or the requirement to abide by certain obligations concerning minimum standards of quality, services or coverage have considerably reduced the margins of activity and increased the need for carriers to invest or find new financial resources to implement their business plans.

This regulation obliged operators to facilitate the introduction into the market of mobile virtual operators (MVNO). MVNO do not have access to the spectrum but sell services with marque, network code, and SIM cards on the network of another carrier. The first MVNO was Virgin Mobile in the UK, the result of a 1999 alliance between One2One and the Virgin Group. In 2000, there were 68 MVNO concentrated in Denmark, Germany, Luxembourg, Finland, Sweden, and the UK. Nevertheless, argument continues in Europe over whether operators with their own networks should open them up or not to these MVNO, meaning that in practice the debate, or the contradiction, between protectionism and the liberalization of the sector has not been resolved.

In Spain in particular, MVNO turnover grew significantly in 2013, 16.9 % up over the previous year (CNMC 2014a). In that same year, 130-plus countries already had a national broadband plan (Broadband Commission 2013) although the degree of implementation of such plans varied greatly depending on the member states (European Commission 2014).

In Europe several essential regulatory features are still awaiting resolution. The first has to do with harmonizing use of the radio spectrum, which determines that a European 3G operator has to function in different bands of the spectrum depending on the country or even region. The second issue has to do with the neutrality of the network and the regulatory asymmetry that occurs when a traditionally State-protected, carrier-dominated telecommunications sector converges with the computer sector, which from the beginning has been free, open to competition, one where new business models appear which groups are itching to develop and exploit. Privacy and Internet security are two other unresolved issues (Acquisti et al. 2013; Pérez and Badía 2012). Neutrality is another. Network neutrality generates major debate on the net between rival interest groups operating in the ecosystem, i.e., individual users or consumers, telephone operators, service, content or app providers, and the big Internet groups.

According to the principle of network neutrality (Wu 2003), all contents circulating on the Internet must receive equal treatment and should not be discriminated against on grounds of origin, use, or application. The concept of neutrality implies universal service (equal access for all individuals) and public transport service (common carriage), which refers to equality of treatment for all contents circulating on the web (Fernández 2014). In other words, users should not find their right to freedom of expression, nor their access to certain contents, sites, or information being limited without their consent. There would seem to a consensus to defend transparency, nondiscrimination of traffic, nonblocking or degradation of access to contents, services, applications, or terminals, and respect for free competition between network and content providers.

Although, from the viewpoint of technology users, producers of contents, applications, and services it would not make much sense to establish limits on the traffic of information on Internet, carriers and network providers defend their rights to discriminate between different speeds or tariffs depending on the services offered, (Fernández 2014) as an incentive to monetize its commitment to developing networks and infrastructures. Internet access providers could discriminate and give priority to certain packages of contents if they need faster or slower carriage, just as they might do so in the case of service users, which means they really would be manipulating the traffic of contents or its quality and favoring certain interests or groups. The capacity to favor certain users or enter agreements with particular providers would give rise to genuinely anticompetitive practices that might put a brake on the generation of contents and progress in the sector generally. Arguments for network neutrality seek to avoid such abusive practices and inequalities in traffic (Califano 2013).

Governments should intervene to guarantee universal access to Internet, support investments in new networks, the protection of fundamental rights and privacy and the confidentiality of information. By doing this, they would also assure the economic sustainability of the network, and user protection. This latter aspect is a key issue for regulation in view of the huge market power involved in the concentration in ownership structure and contents that affects Internet. Google and Facebook are probably the most significant exponents in this regard.[3]

---

[3]The European Commission recently accused Google of abusing its dominant position, suspecting the corporation of giving priority to its own contents in searches and obliging users to use its services and tools to manage their publicity and marketing. Although these are subtle and as yet unresolved issues, they evidently require a regulatory framework generating confidence in an environment of change and radical uncertainty like Internet. Such a framework would prevent abusive practices and guarantee universal access to the net.

So the debate is on and open, with very different hard-to-reconcile interests in play. However, apart from a few initiatives in the Netherlands and some Latin American countries, most countries have not formulated specific laws to regulate neutrality on the net. In some cases, some general principles do exist, as in the United States, established after wrangles with specific provider companies. Users must have the freedom to connect their devices, execute applications, receive the contents packages they want, and obtain relevant information on the service plan hired (i.e., transparency principle). These are what are known as the "four net freedoms" (Federal Communications Commission 2005).

In Europe, there is no clear position on net neutrality, and arguments are still raging over the draft regulations for the European digital single market. In 2010 the Body of European Regulators for Electronic Communications (BEREC) was set up to help develop the internal telecommunications market, encourage the exchange of experiences, and help national bodies on technical issues in applying the sector's European regulations (Directive 2002/21/CE of the European Parliament and Council). In practice, the European Commission proposed that national regulators should oversee service access quality and the transparency of information for users. However, member countries have not issued any specific legislation on this point.

Protecting privacy and the confidentiality of information and security on the net are other issues already mentioned that spark controversy about the Internet. An enormous amount of personal information is available in digital format and this enables firms to design a broad range of economically very valuable services adapted to consumer needs, preferences, and desires. Some authors draw attention to the profits this entails for corporations commercially exploiting such information resources when launching new businesses. They also point to the fact that the consumer is not always aware of the real use made of the data and personal information they give to the big groups (Acquisti 2010; Bijlsma et al. 2014).

The major Internet groups (Google, Facebook, Twitter) have a virtually unlimited volume of personal information available thanks to the free services they have offered through access to social networks, online search engines, e-mail, data storage, and so on. Data traffic via mobile technology is dominated by social networks. In 2014, Facebook and YouTube accounted for 16 % and 15 %, respectively, of all data traffic in the US; in Spain, the figures were similar: Facebook accounted for 20 % and YouTube 10 % (Ericcson 2015). Mobile devices and access to ultrafast networks have brought Internet even closer to the population at large.

In Europe much progress has been made on privacy and data treatment regulation and this has facilitated a legal framework for the protection of ordinary citizens. Although this makes it more difficult to launch new services on the market and places more restrictions on the treatment of data, which might give rise to legal claims and liabilities, technology is advancing much faster and now allows data transfer and storage in a global environment between a host of geographical areas subject to differing regulatory systems.

In the US regulation affects issues that basically have to do with specific sectors dealing with more sensitive data, be it financial, health-related, or associated with minors. Corporations have, though, been given greater freedom to encourage the development of the digital economy (Alonso et al. 2014).

In the early years of the Internet, governments were expected not to interfere (Baird 2002). Discussion between the players involved has, however, gradually grown about the management and control of net resources and several Internet Governance forums have appeared (Pérez and Olmos 2008). In any case, this is an area awaiting development. It also harbors issues related to the protection of intellectual property or the security of information flows. Relations between individuals, corporations, institutions, physical objects, and social networks through the new communication processes and terminals generate a major risk of sweeping information contents disclosure. In this case, it is essential to ensure that consumer rights are protected and guaranteed, whilst also assuring the in-house security and maintenance of confidential information.

Summing up, the sector has enormous business potential, with new business opportunities and new ways of reorganizing both business activities and private life. However, major investments in networks are still required, as Europe has fallen behind its competitors in the US and Japan. A regulatory framework is also needed to guarantee that everyone has untrammeled access to communications and market competition, and to cover essential issues like the protection of privacy and copyright and data traffic security.

4 Regulations in Spain

In the previous section we have provided an overall perspective on regulations at the international level. In this section, and given our geographical location, we are particularly interested in focusing on the regulations that govern the Internet ecosystem and the provision of telecom services in Spain. Spanish law gives the state the power to manage, plan, and control the radio spectrum. Blocks of the digital dividend spectrum did not become available until January 2015. This is a fundamental strategic resource that determines the deployment of 4G technology and coverage of the ultrafast mobile broadband to 98 % of the population in the country and the fulfillment of the objectives of the Digital Agenda for Europe 2020 (COM 2010).

Efforts to develop an information society in Spain began in early 2004 with the move to implement digital terrestrial television. However, the entire process has been marked by major political dissension that has often led to open competency conflicts between administrations. Such conflicts have delayed and hindered progress on liberalizing the telecommunications sector. Concentration within the sector has proved to be another obstacle (Table 2). In 2012, 90 % of the mobile broadband market was shared by just four major carriers: Movistar, Vodafone, Orange, and Yoigo.

Operators	Market share (%)
Movistar	34.81
Vodafone	27.4
Orange	19.8
Yoigo	9.4

Table 2 Market share of mobile broadband in Spain (2012)

Source CNMC (2014b)

This structure has had consequences for the procedure of awarding and allocating the spectrum via tenders. In 2011 the digital dividend (i.e., when analog TV channels were closed to be replaced by digital channels giving rise to the liberalization of many spectrum resources) put 70 % more spectrum resources on the market than had until then existed. Eight licenses in the 800 to 900 MHz band were awarded. However, given the amount of spectrum available and the very low number of companies tendering (Telefonica, Vodafone, and Orange), the licenses were in fact shared out between the big three. Consequently, there was no chance of new competitors coming onto the market. Furthermore, revenues obtained by the state were lower than expected. In Germany however tenders of a similar design generated revenues of some €4300 million as opposed to the 1647 million collected in Spain (Ganuza and Viecens 2014).

License allocation procedure is vital for operators to be able to expand in the 4G business, to obtain advertising and information space. It is also essential for financing Internet companies and, from the government viewpoint, is a source of revenue. However, the oligopolistic structure of the sector in general throughout Europe, and in Spain in particular, has put a brake on the entry of new competitors and earnings.[4]

With regard to investment in ultrafast networks, business concentration in a few firms may favor the deployment of new networks and innovation, although there are some contradictory features that have not helped toward modernizing the telecommunications sector in Spain. Specifically, the number of institutions with responsibilities in sector regulation is so high as to cause problems of competency, lack of consensus, and the harmonization of decisions:

1. The Commission for the Telecommunications Market (CMT) is an independent regulatory body. In practice however, many of its regulatory powers are in the hands of the Secretary of State for Telecommunications and the Information Society, which is not independent and is therefore conditioned by the government's interests.

2. The Ministry of Industry manages the radio spectrum or range of frequencies usable for communications by mobile telephone.

---

[4]In 2000, Spain allocated via tender four Universal Mobile Telecommunication System (UMTS) licenses, the standard for 3G, to Telefónica Móviles, Vodafone, Amena, and Xfera (subsequently Yoigo). Spain was, after Finland, the second country in Europe to do so.

3. The Ministry of Science & Technology authorizes the tariffs for the services of the dominant operator.
4. The Ministry of the Economy safeguards competition.
5. The Ministry of the Treasury charges the reserve fee for private use of the public radio spectrum domain.
6. The European Commission questions regulatory decisions.

It is fair to say that the liberalization of the telecommunications sector, which began in 1996 in Spain, has generally been very positive as far as tariff reductions, innovation in services, bringing the new technologies to the population at large, and deploying the new broadband networks are concerned. But the mobile telephone sector in Spain is basically a restricted oligopoly in which operators have entered gradually, thus allowing the already established carriers to maintain their advantage because of the costs involved in changing operator. In other European countries however, including Germany, Denmark, Italy, France, the UK, Greece, and Portugal, licenses were simultaneous and market shares are much more uniform.

Other important features include a lack of coordination and the conflicts of powers in Spain with local administrations, mostly over the social impact of the effects of mobile telephones on health. Such concerns led to many permits for installing antennas, and for creating the infrastructures, being frozen, which, with the crisis affecting the sector from 2000 to 2002, paralyzed investments in new networks.

In 2001, around the time Telefonica and Amena were awarded two new licenses, MVNO were created (although they did not actually arrive until 2008) to enable fixed landline and cable operators to sell mobile telephones using existing operator networks. Yoigo agreed with Telefonica to use the latter's 2G and 3G networks and Euskaltel did the same with Vodafone. Although these operators' market shares are not particularly high, they are growing and inject dynamism and competition in the sector.

From this angle, it is clear that sectoral policy had some pretty contradictory objectives. While trying to increase social well-being by improving competition, such policy has also protected the sector's industry through measures that tended to maintain the profitability of existing carriers and assure their growth. This can be seen in the rigidity of low prices or the refusal to change the oligopolistic structure, despite encouraging the entry of MVNO and the portability policies that make it easier for users to change operator (Calzada and Estruch 2011, 2013).[5]

In 2005 the Avanza Plan was approved and, subsequently, to give the plan continuity and take a further step toward the knowledge society in the framework of the European Digital Agenda, the Avanza Plan 2's Strategy 2011–2015 was introduced, defining five essential strategic lines: infrastructures, confidence and

---

[5]Initiatives launched in Europe provided the frame of reference for the development of ICT in Spain. The five-year i2010 initiative approved in June 2005 was the EU's first coherent global strategy for the development of the Information Society in Europe (COM 2005 229 final).

security, technological training, digital contents and services, and development of ICT (Ministerio de Industria and Turismo y Comercio 2010).

Spain has defined its own Digital Agenda to encourage the deployment of ultra-fast broadband networks, optimize use of the radio spectrum and improve service quality. As noted previously, in Spain sector regulation is unfocused, particularly because a range of public administrations and institutions have certain powers over the telecommunications sector. To facilitate progress in deploying next generation access (NGA) networks, the Spanish General Telecommunications Act (BOE official state gazette, Law 9/2014, 9 May) was recently passed. This new law (i.e., the *Ley General de Telecomunicaciones*) establishes uniform regulation throughout the country by creating mechanisms enabling coordination and cooperation between the state, autonomous (i.e., regional) communities, and local bodies. This should facilitate the deployment of networks, as sector regulation is what determines things like fixing electromagnetic radiation limits without conflicts between administrative levels (Ballestero 2013). The law simplifies the procedures regulating the deployment of infrastructures by substituting the need for regional or local licenses (e.g., license or authorization for installation, operation, or activity or environmental licenses) by more agile mechanisms like statements of responsibility. It also facilitates access to available civil works infrastructures capable of accommodating telecommunication networks, while also establishing the obligation to develop infrastructures in newly developed or built-up areas to protect free competition and environmental sustainability. Likewise, it facilitates the deployment of fixed networks in buildings and on façades and guarantees the right to access ultrafast networks, so local communities will have to allow the deployment of such networks. The law in fact helps mobile telephones to advance by reducing the cost of deploying NGA networks, giving operators the incentive to extend network coverage. They also have a wider margin to pass on such cost reductions to final prices, this benefitting the end user.

The regulation recognizes the importance of genuine competition as an efficient means of bringing pressure to bear on prices, service quality, and innovation, which explains why: (i) it contains measures guaranteeing the production of regular market analyses; (ii) it improves telecommunications users' rights concerning the protection of personal data and privacy, setting out sanctions; (iii) it creates a cross-ministerial commission on radio frequencies and health, to improve environmental safety; and (iv) strengthens the powers of inspection and sanctions held by the Secretary of State for Telecommunications and the Information Society.

The regulation also attempts to encourage flexible use of the spectrum (i.e., share), operator-to-operator agreements on sharing infrastructures and investments, facilitating public support in zones where deployment is limited by lack of profitability. Likewise, the drive to deploy infrastructures must be accompanied by policies designed to stir up demand for broadband services amongst the population at large, companies and the authorities. Initiatives planned in the 2013 Digital Agenda for Spain (Ministerio de Industria and Energía y Turismo 2013) and specific promotion plans spinning off from this Agenda take account of the impact of such measures.

Even so, the deployment of 4G technology in Spain began in 2013, after considerable delay, led by Yoigo (TeliaSonera) and Orange, covering about 60 % of the population. Deployment has been concentrated in the larger cities with download speeds of up to 75 Mbps in the 1800 MHz band (CNMC 2014a). Sluggish use of the band in 800 MHz underscores the fact that 4G coverage is concentrated in specific places and territorial coverage is not homogeneous. The situation is similar in Europe where, although the use of 4G networks is growing, coverage is very heterogeneous depending on the country. In 2013 only eight countries had coverage of more than 50 %.

However, we need to view the sector's progress in light of the forecasts of major growth and of the changes observed in the strategies implemented by provider companies, operators, and the major Internet groups. Below we look at the Internet ecosystem and the epigenetic forces being unleashed in it.

5 Internet Ecosystem and Epigenetic Forces

The mobile Internet ecosystem brings together a host of related activities that shape the advance and development of the Internet. In this environment, layers of technology permit Internet players to interact, innovate, and compete for a market niche (Arlandis and Ciriani 2010). In 2013, the digital ecosystem generated total revenues of €512 billion in the 13 countries that contribute to 70 % of global GDP (Bock et al. 2014). The US generated 35 % of total revenues, China 21 %, Japan 12 % and Germany, the UK, France, Italy, and Spain together 18 %. Players interacting within the ecosystem to determine how the mobile telephone sector evolves include:

- Telephony operators or infrastructure and network providers
- Content, application, and service providers
- Equipment, terminal, and smartphone manufacturers
- Software and operating system developers
- Internet groups
- Users and clients or companies
- Institutions, governments, and regulatory agents

Clearly, there are different technology layers (e.g., networks, terminals, operating systems and applications or contents) and different software and service providers like the major Internet groups (e.g., Apple, Google, Facebook, and Amazon). As a whole, sector growth basically depends on the availability of enough spectrum to transmit an increasingly large flow of information and data, on suitable networks and infrastructures that enable the transmission of information flows, on high-performance terminals with new capacities in terms of computer programs and applications that users will subsequently download on their devices, and on the supply of attractive contents and services. Something like 73 % of the revenues generated in the US and in Europe is produced between the applications and contents sector and the Internet platforms, operating systems and terminals sector. Prospects for growth of the ecosystem as a whole suggest that the 2013 revenues will have doubled by 2017.

The interesting thing about the epigenetic approach is that it enables us to observe changes in the environment from a dynamic perspective (see Chapter "Introducing an Epigenetic Approach for the Study of Internet Industry Groups"). We can detect strategic behavior on the part of Internet players that does not respond to their genetic identity or DNA and which radically transform their traditional specialization. One intriguing feature here is the coming launch by Google (not, in principle at least, a hardware producer) of Chromebit, which converts any screen with HDMI into a computer with web access. Google will shortly be competing with smartphone producers and with low-range computer markets such as Microsoft's and will be looking to win more user net connection time, thus earning space to place its services and advertising. Also detectable are movements by carriers as they begin to develop operating systems or get into the device marketing sector, sometimes with their own-brand products and accessories for smartphones. In Spain, terminal sales through operators currently account for 75 % of the total.

The future of the sector would basically seem to lie with the contents and services and the apps and software businesses where added value is really generated. A 2014 report by the National Telecommunications & Information Society Observatory in Spain identified tendencies shaping future home and individual consumption habits and the way to restructure and reorganize businesses and economic activities.

To begin with, there is the IoT. This is a new area for the creation of communication infrastructures which will enable process optimization through savings in costs and greater economic efficiency, while also facilitating greater well-being and quality of life for the population at large. The IoT involves the possibility of connecting the real and virtual worlds through remote information sensors that connect all types of objects, machines, and even vehicles to Internet. It is a process that combines people, information, and things to create intelligent control systems with a host of applications ranging from healthcare, education, traffic control, driverless vehicles and cars to smart public lighting, advertising, marketing, leisure, culture, smart buildings, and water and energy management. It is not only a question of the enormous potential for innovation in services or processes; the potential is there for product innovation too. Wearables have major market potential as regards new product launches: interactive bracelets, smart watches and, in general, devices wearable on clothes that help to improve people's well-being in healthcare-, leisure-, and sports-related activities.

Generally speaking, people will be using mobiles increasingly to access all kinds of services associated with healthcare, entertainment, social life, transport, e-payment systems, or even when communicating with the authorities. This is what inflates the potential for growth of both e-commerce and in access to social networks from mobile devices, smartphones, and tablets. Forecasts suggest that by the year 2020, some 55 % of global data traffic via mobile technology will be video contents (Ericsson 2015).

Data centers in the cloud also have considerable potential. Using mobile devices, users can access a number of data storage, contents, and services systems housed in the cloud. The "cloud of clouds" (i.e., the possibility of interconnecting private clouds in a global interconnected cloud) in much the same way Internet, in its day, connected individual networks in a global network, will substantially enhance and increase the Internet ecosystem.

What we are seeing is technology converging in a process where mobile technology is a basic vehicle of innovation capable of integrating several technologies linked to the cloud, social networks, intelligent sensors, data analysis, and the development of software for applications. This entails coordinating the research and development decisions of a large number of companies that contribute their complementary innovations to new product developments. A large number of companies with wildly differing profiles and objectives were involved in the development of technologies like 3G mobile telephony. Developers of components for mobile devices coexist alongside businesses providing the operational infrastructure and others manufacturing the actual devices, whether mobile telephones or tablets (Llobet 2014: 51). In these cases, the need to coordinate the activity of so many companies led to forums and standard setting organizations. Such forums helped in developing 2G and 3G. At present, the host forum for discussion and debate on the technology for mobile telephones and specifically 4G is known as the European Telecommunications Standards Institute.

Companies use standards to define the features new products must comply with, so as to make the best possible use of the innovations they include. These standards are, therefore, the result of several years of technical discussions between companies about how best to implement specifications in the best possible way and how to overcome the challenges posed by the new technologies. With mobile telephony, for instance, the licenses for more than a thousand patents are said to be required to produce a device that complies with the specifications of the third generation standard. These patents are called standard-essential patents (Llobet 2014).

Acquiring patents is a key business strategy here. Samsung, principal producer of screens, has invested in networks and infrastructures and is now a leader in 4G-related patents. Other companies like Apple have hardly any patents and may find themselves forced to look for alliances to take on new investments, new markets, or just move forward in technological development. Table 3 lists the leaders in 4G-related patent issues.

Table 3 Leaders in 4G patent-related issues

Samsung	Qualcomm
InterDigital	Nokia Corporation
Ericsson	LG Corp
Motorola Solutions, Inc.	Motorola Mobility Holdings, Inc.
Panasonic Corporation	Sony Corporation
NEC America Inc.	Texas Instruments
Nortel Networks Corporation	Intel Corporation
Harris Corporation	

Source iRunaway (2012: 9)

As opposed to concentration of business activity in the sector, where, in principle, there are no operators, hundreds of small firms hold patents in this area. And not just small firms: research institutes in Asia and the US are also patent holders. Very recent data reveal that despite the concentration of 4G-related essential patents, the three main leaders, Qualcomm, Nokia, and Ericsson hold just 53 % of essential patents (Han 2015).

Although such strategic behavior (which underscores both the major investments made in patents and concentration by the leaders, as well as the interest of smaller organizations in getting more than a foothold in the mobile telephone market) might be seen as an attempt to set up an entrance barrier to limit competition, in practice it has on some occasions led to extortion-like behavior where firms acquire patents (i.e., patent trolls) with the object of demanding license payments with no other interest in research or development, a move that is obviously not designed to encourage innovation. In other words, the war of patents between major groups (Gómez-Uranga et al. 2014) in the mobile technology sector has led to a very lucrative market. It has, however, also become a major source of legal conflicts linked to negotiations over how much licenses should cost, patent infractions, and abusive license conditions (Llobet 2014), forcing competition policy authorities to intervene.

In the following section, we look specifically at carriers' market strategies designed to do something about the uncertainty about the need to make new investments, guarantee technological innovation, and satisfy the need for new services and contents with greater value added in the net. These forces have given rise to not a few alliances, mergers, and acquisitions of companies within the ecosystem as they shape the sector's future.

6 The Carriers' Role

A major challenge for operators now is to maintain growth in revenues, which have already begun to stagnate and even to fall in some activities. A recent report by Little (2015), which analyzed upwards of 560 comparable tariffs for voice and data in 16 countries, found that in most, and particularly European countries (Germany, France, Denmark, Ireland, Italy, Spain, and the UK), there was a clear tendency toward price reductions.

With the introduction of 3G and, in particular, of 4G, business is evolving so fast that it is actually putting operators at risk. Unless they adapt their routines to the epigenetic changes now occurring (see Chapter "Epigenetic Economics Dynamics in the Internet Ecosystem" in this book by Zabala-Iturriagagoitia et al.), they may struggle to survive. The Internet ecosystem operators work in is clearly subject to the Red Queen effect[6] because they are operating in an ever-evolving,

---

[6]The Red Queen theory is a concept introduced by the evolution biologist Van Valen (1973) to explain the extinction of families of organisms in a given time period. It is also applied to the economy to explain the behavior of companies when trying to keep abreast of the competition.

highly changeable system that obliges them to improve continuously just to keep in step with the system with which they are coevolving.

Carrier business has traditionally concentrated on voice, but in a few short years has essentially shifted to focus on data. In this ecosystem corporations like Google, Apple, or Facebook are forcing operators to evolve toward information-centered business if they want to maintain their position. Information is more valuable than data as information is a contextualized datum. This profound change is radically altering consumer behavior; consumers now show less loyalty to an operator offering minutes of voice and gigabytes of data only, because of the nondifferentiation of the product: all bytes are equal, just as all the minutes of calls with which they tariff their client base are equal. An operator basing its approach on the minutes of calls and number of SMS messages is basing its competitive capacity exclusively on price, by offering an undifferentiated product. In this sense at least it would be very similar to a utility that bases its business on delivering kilowatts. This is basically what telephony operators have been doing: working a highly regulated business, as we have seen, which in most cases has its origins in former state monopolies, some dating back to the early twentieth century and which involved heavy investment in fixed asset in the deployment of the network, investments that were easily recouped thanks to major entrance barriers and very low variable costs. This lasted for decades and, throughout that time, operators worked to routines based on this favorable environment. The value of such routines, however, has fallen considerably in the new environment. The working environment has evolved too fast for companies accustomed to competing with very few other companies, accustomed to them being the biggest and therefore, to bringing major muscle power to the negotiating table. For this environment they developed complex routines based on improving efficiency in data transportation, on process efficiency and control, making them rigid, highly bureaucratized organizations. This comes through clearly in their client relations: complex, highly standardized contracts, with rigid, penalty-ridden rules but lacking other routines and powers required to operate in a new and, in principle, rather hostile ecosystem.

In developed countries, the mobile penetration rate is already high, and the market may be considered mature, increasing competition on prices. In Spain, for instance, the penetration rate maxed in December 2009, at 114.3 %, and although the subsequent drop is almost certainly attributable to the economic crisis, it would seem to have peaked. Things are similar to Spain in most other European countries, where penetration rates are upwards of 100 % of the population.[7] Competition between Over the Top (OTT) companies, with offer free voice and messages, negatively affects voice and data revenues, as Fig. 3 and Fig. 4 show for Spanish data, where income for voice and SMS has fallen steadily since 2007.

---

[7]According to World Bank, in the European Union the penetration rate was 125 % in 2013, actually exceeding 150 % in some countries, including Latvia, Finland, Estonia, Italy, and Austria.

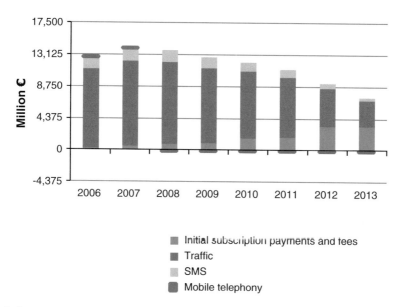

Fig. 3 Revenues of mobile telephony in Spain. *Source* Own elaboration based on data from the Spanish national telecommunications market commission

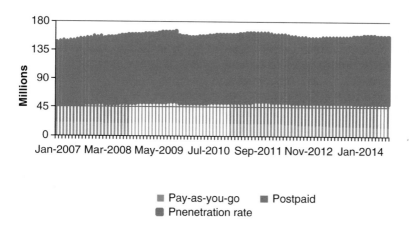

Fig. 4 Mobile lines in Spain. *Source* Authors, from CNMC (2014a, b)

Besides the fall in income, also in the mix is the fact that carriers have to keep on investing in infrastructure to improve the network and adapt it to LTE-Advanced (4G) technology, even though the penetration rate is still fairly low. During the 2007–2014 period they invested €155 billion in Europe on developing network infrastructure and are expected to invest a further €170 billion in the 2015–2020 period. These investments will be used to increase network capacity

Table 4 Market share of
mobile carriers in the USA

	2003 (%)	2009 (%)	2012 (%)
Top 3	53	79	80
Top 4	63	91	93

to cover increased demand in data traffic, and to expand LTE-Advanced coverage (GSMA Intelligence 2014a). Operators globally are expected to invest $1.7 trillion between 2014 and 2020, most of which will be spent on deploying 4G networks (GSMA Intelligence 2014b). Given the sheer amount involved, operator strategy will focus on how to generate the greatest profit possible from the 3G and 4G networks, by developing new services and tariff models.

European operators of course face the added problem of regulation unfavorable to the kind of market concentration found in other regions, basically in the Pacific (China and Japan) and North America (Table 4). In a world of strong economies of scale, this may well prove a tough competitive disadvantage to overcome. In the USA, a major process of mergers and acquisitions has helped to consolidate the sector. In 2003, the three leading mobile operators (Verizon, AT&T, and Cingular) held 53 % of the market, Verizon leading the market with 24 %. In 2012, the first three had 80 %, with AT&T the new leader.

Something similar has also happened in Spain: the top three operators (Movistar-Telefonica, Orange, and Vodafone) share most of the market (84.6 % in December 2012 and 77.1 % in December 2014). Unlike the USA, however, they are actually losing share over time, following the entrance of virtual operators[8] which have gone from a market share of 9.1 % in 2012 to 16.2 % in 2014, owing to the obligation, imposed on the mobile network operators (MNO) by the Spanish national telecommunications market commission to hire network out to MVNO setting up in the market, to favor price competition.

Taking European countries separately, we find a similar situation to the USA, in which the leading three operators take the largest share of the total market. However, the leading three are not usually the same in each one. In European countries as a whole, there are more than 50 separate independent operators occupying one of the first three places in market share in any country.[9] In the UK (BT, Three, and Vodafone), in France (Orange, SFR, and Bouyguess Telecom), in Germany (Deutsche Telekom, Vodafone, E-Plus) and in the Netherlands the leader is TeliaSonera. In short, in European countries we have a fragmented market, in which most operators have a very tough battle to achieve leadership in a number of markets (Deutsche Telekom, Orange, Vodafone, Telefonica, and TeliaSonera do

---

[8]In early 2015 in Spain 47 MVNO were operating; 5 of them were using the Telefonica network, 7 the Vodafone network and 35 using Orange's.

[9]Apart from the big names (Vodafone, Telefonica, Orange, BT, Three, MTN, and TeliaSonera) other operators, including Cosmote, Wind, Siminn, Eircom, LMT, Post, Moldcell, KPN, AL Betelecom, MTS, Telekom Austria, Turkcell, Belgacom, Telenor, Tele2, SFR, and Bouyguess Telecom are amongst the first three operators in some European markets.

so partially). This is because in Europe diversified operators are no more profitable than those that operate in just one market. Multiple regulators, the fact that the band spectrum is put out to tender country by country with different modes of tenders in each one, and the difficulty in achieving economies of scale in marketing, owing to the sheer number of languages, all favor fragmentation. Nor is it coincidence that in most cases the leaders are state-owned operators (Deutsche Telekom in Germany, British Telecom in the UK, Orange in France, TeliaSonera in Sweden and Finland, Cosmote in Greece, Telenor in Norway, etc.) or former state-owned carriers (for instance, Telefonica in Spain or AT1 Telekom Austria in Austria),[10] which underscores the advantages for first movers and disadvantages in costs for challenger firms. State participation in shareholdings hinders cross-border mergers and acquisitions, as few governments are going to be happy about relinquishing control to foreign operators.

All these factors negatively affect income and Average Revenue per User (ARPU), a key variable in an operator's profitability and growth potential (Network Strategies 2013). Operators are hoping that the client migration from 2G and 3G to 4G, a move driven by higher data consumption, will produce a significant increase in their ARPU in the future, an increase that will put them back on the track to growth in revenues, as noted at the beginning of the chapter, but at present the 4G-LTE penetration rate is not particularly high globally. Only in South Korea and Japan are they really high, at 70 and 45 %, respectively, in 2014 and these account for 20 % of all 4G lines the world over (Ericsson 2015). The United States, which is ahead of Europe in deploying 4G, had a penetration rate of just 19 % in 2013, while Sweden, the spearhead in Europe, had 15 % and major countries like the UK and Germany a modest 5 and 3 %, respectively.[11] It is true, though, that reports unanimously agree on major growth for 4G globally. By 2020, the forecasts point to 3.7 billion subscriptions, when in 2014 there were only 500. This means annual growth of 40 %. Most LTE lines are concentrated in the Asian Pacific region. Some 1100 million 4G subscribers are expected to materialize in China alone.

Although 4G increases the potential for differentiation and sales of other additional products, the truth is that its positive effects on ARPU are still pretty weak. A 2013 report by consultants Network Strategies, looking at operators in Sweden, Finland, Denmark, Norway, Australia, and the US, suggests that ARPU for 2009–2013 in all of them continued to fall; that being first movers in introducing 4G has not given them any advantage in terms of ARPU with respect to the followers and that in all of them, except in the three Finnish operators, the introduction of LTE

---

[10]Telefonica was a state-owned firm until it was privatized in 1997. Telekom Austria was privatized in 2000. It was bought by America Movil (59.7 % of the capital) in 2014.

[11]Europe is in fact aware of the importance to the economy of deploying the 4G-LTE network and major growth is expected in penetration rates. GSMA Intelligence (2014a) expects the rate to rise from 10 % as of early 2015 to a situation where, by 2020, more than half of all mobile connections will be 4G.

Advance has not meant a price premium. However, these results do not invalidate the hypothesis that there is a positive correlation between introducing 4G technology and ARPU. It is normal for positive effects not to be appreciated for a time, as three other factors have also to be present: coverage needs to be extended to large areas of the country or of the population, terminals for this technology have to be deployed and clients have to migrate from 2G and 3G to 4G (many do not a first see the need to do so).

All these factors have to date prompted analysts to consider operators as utilities, because they do not expect them to obtain growth rates above the GDP. This view has some major consequences for operator strategies: if the market does not expect major growth in revenues, it will prefer profits to be shared out as dividends rather than be plowed back into the business, penalizing companies that do not give dividends in their market value. This is what happened to Italia Telecom (Fig. 5): in 2004 it paid 74 % of profits out as dividends, while in 2013 it paid out just 13 %. The process became particularly noticeable from 2009 largely because of the need to invest in 4G, but as dividends fell, so did its market value over its book value, which went from 2.62 to 0.76. In general, operators have maintained stable high dividend rates (Table 5). AT&T, for instance, has a payout ratio of 75 %, TeliaSonera maintained an average payout ratio of 86.8 % in the 2004–2013 period, Deutsche Telekom, 116.2 % in 2014, and Telefonica 110.75 % in 2011.[12] These are high values when compared to corporations like Apple, which in the same period returned an average payout ratio of 26 %, despite having major liquid assets; Microsoft had a 30 % ratio and—for the moment—Google retains the huge profits it generates. The market does not punish these firms for not remunerating shareholders, as their higher payout and market to book value demonstrate.

In short, operators find this environment hard to manage. Companies that continue to give high payouts are obliged to use more debt to carry out mergers and acquisitions and to finance the high capital expenditures required for the deployment of 4G and fiber. The problem is that companies already have high levels of debt. Indeed, they often have to disinvest in other markets to be able to carry through acquisitions while avoiding getting further into debt. Telefonica used its disinvestment in O2 to strengthen its position in the German market (E-Plus) and to reduce leverage, bringing down net debt of 2.73 times to 2.13 times its operative income before depreciation and amortization. To put it succinctly, carriers face major changes that are forcing them to rethink their business model, as revenues from their traditional business continue to decline and the possibility of offering differentiated products diminishes. Voice and message services scarcely provide perceived value to clients and so operators find it difficult not to compete in prices either.

---

[12]In 2013 and 2014 Telefonica paid no dividends out to reduce debt.

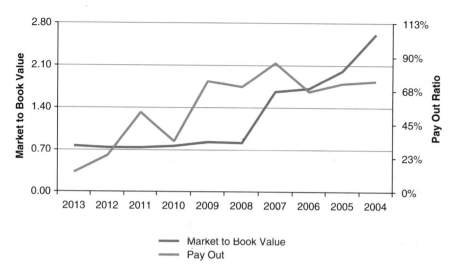

Fig. 5 Italia Telecom. *Source* Prepared by authors using data from Italia telecom financial reports

Table 5 Payout ratio and market to book value in selected carriers and OTT companies (2014)

Carrier	Payout ratio (%)	Market to book value	Over the top firms	Payout ratio (%)	Market to book value
AT&T	75	2.08	Apple	26	5.26
Telefonica	0.00	2.6	Facebook	0	6.14
DT	116.16	2.6	Google	0	4.28
TeliaSonera	86.8	1.9	Microsoft	30	3.96
Vodafone	30.50	0.91			

Source Authors' elaboration

So far things have looked rather somber for operators, in a scenario where they make the investments and the OTT capitalize profits the network generates, as the former have not been able to reap the revenues thus engendered. However, the new 4G-LTE makes rewriting the rules of the game a possibility, thanks to its disruptive potential and to the fact that to establish a competitive advantage a firm needs to be capable of forming or penetrating LTE-A-related ecosystems.

To recover profitability and growth potential, besides increasing operational efficiency, the strategic approach of the operators is based on rethinking their relationship with the OTT and on entering the digital economy and transforming themselves into a Telco 2.0.

6.1 Operators and OTT

OTT services use Internet as a platform to reach the end user. There is something *winner-takes-all* about them, as for many variable costs are practically nonexistent, with most fixed costs being sunk costs. They are generally subject to network effect, which tends to give them first-mover advantages (Lieberman and Montgomery 1988). First movers can develop isolating mechanisms to protect themselves from imitators by generating entrepreneur revenues for themselves (Rumelt 1987). The cognitive switching costs created by forming habits in consumers (Schmalensee 1982), particularly present in, for instance, applications for social networks or reputational advantages (Carpenter and Nakamoto 1989) are some of the isolating mechanisms OTTs can set up.

Companies use the Internet in many fields as a platform from which to offer its services, thanks to spectacular technological developments. The LTE-Advanced technology is going to accelerate this trend, because it opens up enormous possibilities for digitalizing services to the economy even further. Operator profitability has been particularly hard hit by OTT communication over IP apps for voice like Skype or Viber, and message apps like WhatsApp because they provide these services using freemium as their business mode, which means they are directly attacking traditional carrier core revenues, with the distinctive feature that the OTT have not had to bear network deployment or maintenance costs. As Fig. 6 shows, SMS revenues fell 60 % in two years in Spain, and estimates are that between 2012 and 2017, carriers would lose $3000 billion in income from their message services as a result of OTT activity. The OTT also have a major impact on video services: Youtube, Netflix, Waki TV, Amazon Instant Video, etc., owing to the increase in data traffic generated over the network, are putting pressure on operators to invest to increase transit capacity.

Carriers are aware that simply being a distributor of information stops them from appropriating the huge value added currently being generated in the Internet ecosystem. The evolution of market value is an indicator of this. In the 2009–2013

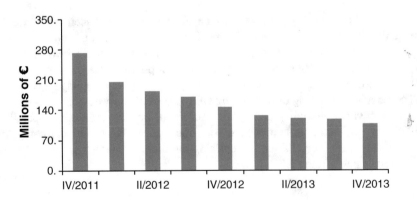

Fig. 6 Revenues of SMS in Spain. *Source* Own elaboration based on CNMC (2014a, b)

period, the market value of the world's ten largest operators[13] went from $967 billion to $1057 billion, a 9.3 % increase, while, in the same period, the largest OTT[14] went from a market value of $776 billion to $1.88 trillion, an increase of 142.5 % (Bock et al. 2014). And yet this value added is being created thanks to the technological developments in speed and information transport capability over the network, where carriers are the prime players. Ganuza and Viecens (2013) reckon that high-speed networks are going to facilitate a reallocation of revenues between players in the value chain. In particular, they will migrate to contents providers as they do not need to negotiate with carriers to gain access to users, doing so directly through a web page in a de-intermediation process.

The operators' first reaction was to tackle the OTT by applying to the authorities for regulatory changes, such as acting on the neutrality principle, as noted above, and by—unsuccessfully—requesting permission to pass on to the OTT the cost generated for them by the increase in data traffic (Telefonica, KPG, and Italia Telecom have all applied to the European Commission for this). At the 2013 World Mobile Congress, Telefonica's CEO declared, in calling for symmetrical regulation for the other players: "Investing more on terminals than networks, or paying huge amounts in fees and spectrum, while being called on to speed up the deployment of new generation networks, is simply not sustainable." Carriers have also tried to oppose the OTT by creating equivalent message services. In 2012 Telefonica launched an instant message service to fend off WhatsApp and Line, but only managed 1 million downloads, as opposed to the 300 million-plus logged by WhatsApp or Line's 200 million. In 2006 it launched Pixbox, an online music and movie store, to compete with Netflix, Apple, and Waki, but it closed down in 2011. Competing with the OTT, which have already taken full advantage of the major network externalities, is a Herculean task.

On other occasions, however, operators have sought alliances with OTTs in a bid to appropriate to themselves revenues in the Internet ecosystem. The network has value to the extent that it has contents, and content providers need high-capacity networks to be able to offer them faster and more widely: carriers and OTT need each other. By way of example, Deutsche Telecom, KPN, Telefonica, and Vodafone have signed agreements with Spotify, Vodafone and Three with Skype, AT&T with Beads. Operators seek to differentiate themselves from other carriers and strengthen loyalty in their client base, thus reducing the churn, or attrition rate.

Carrier strategy for appropriating to themselves the value they generate should involve a shift in perspective as regards perceptions of the network. Rather than seeing it as a set of physical and technological resources, carriers need to conceive the network as a set of intangible resources and capabilities made up of thousands of routines that enable them to make of the network a business facilitator. They should also discover and package the experiences of their remarkably broad client bases: operators can reel in part of the value of the network by creating value for the OTT.

---

[13]China Mobile, AT&T, Telefonica, Vodafone, Verizon, France Telecom, Deutsche Telecom, NTT, NTT DoCoMo, and America Movil.

[14]Apple, Microsoft, Google, Amazon, Facebook, eBay, Yahoo, Baidu, Tencent.

Friedrich et al. (2013) also understand that carrier strategy for appropriating some network value should avoid confrontations with the OTT and look to cooperate by taking advantage of the resources they already have: they can provide ubiquitous connectivity, they can use their colossal client base as a distribution network that they could offer to selected OTTs.[15] They can also expand toward adjacent businesses using their technological know-how and taking advantage of the seemingly limitless quantity of information circulating around their networks.

6.2 Carriers Entering the Digital Economy

As we have already noted, with 4G technology the carriers' objective is to reposition themselves in the value chain and get into areas of value added to return to growth and profitability. In developed countries, the combination of smartphones carrying 4G technology and the deployment of ultrafast networks is ushering in profound sociological and behavioral changes. It is radically altering the ways we move around, buy things, and look after our health. In this ecosystem that revolves around mobile devices and the cloud, relations between players are going to be more complex, varied, and open, being in general more collaborative than in the past. In such a world, it will be vital to have the capacity to lead projects that attract the players necessary to create microecosystems that satisfy the specific needs of a broad group of consumers who will have to be provided with personalized solutions to their requirements. Only global carriers will have the capacity to do so.

To enter the digital business world, speed is of the essence, given the possibility of network economies springing up and elevated costs for client-base change. Critical masses are high in such conditions and as we have seen in app stores, arriving late means you do not get in. Vodafone tried without success with the Vodafone 360 platform, but attracting a sufficient number of app developers to make it as attractive as Apple Store or Google Store proved too difficult.

As we have already said, carriers do not have the time required to develop the routines to build the competencies needed to achieve a competitive advantage in the sector. To do so, the option chosen by carriers successfully repositioning in the value chain of the digital ecosystem, and for which the term Telco 2.0 was coined, is to develop networks of alliances.

A lot of companies are going digital, so we might expect to encounter a broad range of diversification strategies in the Telco 2.0. However, a certain pattern is discernible in the strategic movements they have made. They have opted to enter adjacent sectors, which are the ones where they can best take advantage of their routines and capitalize on the competencies developed when negotiating alliances with their partners. They are sectors in which experience in developing

---

[15] An example of this is Telefonica's agreement with the *Circulo de Lectores* book club.

technological solutions while installing networks is important, as is the ability to access a huge client base, and where the network is needed to provide service. In view of all this, the great majority of carriers have opted for entertainment, healthcare, finance, and the IoT.

6.3 Carriers and the Entertainment Industry

One of the most important forces to have appeared in the Internet ecosystem and which is modifying carrier business models is the convergence of television and the telecommunication industry. Improvements in network quality and capacity as a consequence of 4G technology and the ultrafast Internet of the kind achieved with fiber to the home with speeds of 1 Gb per second is accelerating a momentum that began with 3G.[16] This drive quickly prompted carriers to make incursions in the business. Indeed, entering pay TV may be considered the most natural expansion for the telecommunications business because high-definition television is the main generator of high-performance network demand, in fiber and 4G alike.

Given the range of routines required, this movement has been made through acquisitions (e.g., Direct TV by AT&T,[17] AOL by Verizon, Canal + España by Telefonica, EE by British Telecom, Turun Kaapelitelevisio by TeliaSonera, Jazztel by Orange, etc.) or through alliances (e.g., TeliaSonera signed a contents distribution agreement with HBO in 2013 and another agreement, with Samsung, to develop Internet Protocol television technology enabling people to watch TV without a set-top box).

Behind this dynamic was the desire to offer the client packaged products, including fiber, Internet, mobile telephony, and television, called quadruple play. Most major carriers have taken up positions in quad play-style service packaging because they see in this strategy a way of making use of the network profitable (Telefonica reckons that fiber clients have an ARPU 1.5 times higher and half the churn as low capacity networks), taking advantage of economies of scope and operating efficiency (e.g., in billing clients thanks to the single invoice), and above all of the potential for differentiation it offers, facilitating more personalized packages that hike the cost of changing. Grzybowski and Liang (2014) examined quad play demand and found that consumer surplus increased with the introduction of quad play packages, and was greater when based on fiber to the home rather than asymmetric digital subscriber line (ADSL). Entering digital pay TV enables companies to sign contents exclusivity contracts with contents providers. So it is a way of defending themselves from OTTs like Netflix.

Where the influence of 4G on the convergence of the entertainment and telecommunications industries is clearest is in the personalized way spectators will

---

[16]In this "epigenetic" drive, Google is one of the players entering the ultrafast Internet business and offers clients à la carte TV contents.

[17]This is the greatest monetary amount of all: $49 billion.

see live televised events. For instance, EE has developed a technology, set to come onto the market in 2016, that will revolutionize the way we view live events: users will use their mobile telephones to choose the TV camera that is recording.[18] The technology was trialed during a soccer match at the Wembley Stadium. Besides being able to choose at all times the part of the pitch they want to see, spectators will be able to repeat a particular move as often as they like before returning to the live broadcast and, thanks to an app downloadable on their telephones, obtain information they want about any player, like the kilometers covered, passes made, fouls committed, etc. In short, 4G makes it possible to offer spectators a much more enhanced experience to the one offered by traditional technologies. But, as is happening in the other sectors that are digitalizing around Internet, this technology requires the participation of a range of companies forming an ecosystem. The BBC, Qualcomm, Huawei, a leading firm specializing in multi-camera recording and on the fly editing (EVS) and a developer of apps for mobiles (Intellicore) have all been involved in the EE initiative.

The convergence of industries prompted rapid mergers, acquisitions, and alliances because, just as in the other adjacent industries that are currently going digital, carriers lack the routines and capabilities to operate in them by themselves. Developing such routines and abilities would take far too long. As speed is so important, a carrier entering the entertainment contents sector through mergers, acquisitions, or alliances with other firms generates a knock-on effect in the other carriers to prevent them being left behind. If the others do not react fast enough, the best pairing options will probably have disappeared. This happened to Yoigo (TeliaSonera) in Spain, after Vodafone and Orange's rapid reaction to Telefonica's movement with the quad play Fusion package, which it accentuated with the acquisition of Canal + España from Mediaset and Prisa. Vodafone acquired Ono, a fiber carrier with TV contents and Orange acquired Jazztel before Yoigo made its move. Now Yoigo's position in mobile telephones is very weak, as it has a bare 6 % market share. With such a low market share, it faces problems in making its investment in 4G profitable, despite having reached agreements with Telefonica to share network wherever Yoigo does not reach. In the UK, movements in packaging have also provoked rapid reactions by competitors. After BT invested £3000 million on deploying fiber to the home networks, and then immediately bought the exclusive broadcasting rights for Champion League soccer games (£900 million), offering its clients exclusive TV contents (BT Sport), plus the acquisition of EE from Orange and DT, Vodafone, Sky, O2, and Three had no option but respond with merger and collaboration agreements. Vodafone attempted to acquire BSkyB in late 2014 to counter BT's presence in pay TV.[19] It was in fact a move the market had been expecting, as Vodafone had already entered the cable TV business in

---

[18]http://ee.co.uk/our-company/newsroom/2015/05/31/EE-trials-live-4G-broadcast-of-FA-Cup-final-at-Wembley-stadium (last access April 15, 2015).

[19]Sky has broadcasting rights on the English Premier League, and after receiving permission from the fair competition authorities to acquire Sky Italia and Sky Germany, to bring them together in Sky Europe, has become Europe's largest pay TV group.

Germany and Spain, also through acquisitions (Kaberl Deutschland and Ono respectively) and also had cash available after its disinvestment in Verizon.[20] Subsequently, Sky and TalkTalk tried for collaboration agreements, and even a merger deal with O2. Vodafone, after failing to acquire Sky, tried with Virgin Media, acquired 2 years previously by Liberty, a cable carrier with which it has explored the possibility of an asset swap. In short, the convergence of contents has sparked a sprint involving mergers, acquisitions and alliances between carriers to avoid being frozen out of the market. Carriers perceived that not being involved in the race for contents convergence would leave them debilitated. In the British market, when O2 was unable to materialize any of the agreements it had explored to enter as a quad play carrier, and as margins in mobile telephones in the market were narrowing and its parent company had high leverage, it decided to abandon the market and accept Three's acquisition offer.

While convergence is the predominant note, not all carriers see it as beneficial. In Three's case, what was at stake was not an acquisition of or merger with a contents company, but the purchase of a mobile carrier with a hefty presence in 4G, to be the leader in the mobile telephone market. Instead of looking to diversify, it plumped for specializing and concentrating its resources in this market. Three CEO David Dyson (Thomas, 2015) was not sure consumers really wanted to buy all their services from a single provider. "It all depends on how aggressive the operator is in pushing discounts on parts of bundles" (ibid). So, in the short term, although this strategy may help an operator gain users and better recoup investments in fiber and 4G networks, the problem lies in whether it will last in the long term if most carriers follow the same strategy, as is now occurring. For some, greater consumer loyalty, lower churn and the synergies obtained through single billing will be sufficiently important to compensate price-war fallout. For others, though, this is not so evident. However, operators may come up against problems of differentiation in the future when all of them offer quad play packages. Entering the contents industry means they will have to develop royalties management routines, something they are not prepared for as yet; savings on costs from single billing for packaged services (utilities are already doing this with gas and electricity) are not clear either because changes in accounting rules (IFRS 15 and FASB Topic 606) for revenues and costs on service contracts are going to force carriers to invest in information technology systems that include particularized information from the contracts. For the moment, surveys in the UK suggest consumers find offers combining Internet, television, and telephone more valuable than separate offers, and CCS Insight predicts that, by 2019, 80 % of homes will be subscribed to some operator package or other.

---

[20]Although in principle this disinvestment may favor the acquisition of other firms, through having the cash available to do so, the downside is usually that it increases the acquisition premiums, usually quite high per se.

6.4 Operators and Financial Services

In the economy's digitalization process, the banking business is at risk of undergoing major reshaping at the hands of the OTT. The bank could actually lose its monopoly on providing financial services. Apple, Google, and Facebook have just started offering money wallet services, and these moves are beginning to generate debate about whether OTT could transform themselves fully into banks or if they are only interested in offering some financial services based on the payment of electronic transactions. It is still too soon to answer this question, as technology is not the only variable in play. Regulation is a capital factor for an organization to be considered a financial intermediary, as is the confidence depositors have in being able to recover their money whenever they want. But core banking activity is based on information. Basically a bank works as an intermediary, taking savers' (depositors') money and loaning it to anyone who requires it. There is a problem of information asymmetry in this process, and minimizing it means incurring in significant transaction costs. The subprime crises are a major consequence of these asymmetries. If information were perfect, banks would not be necessary, as matching those needing savings and those offering them would not be at all costly, and the latter would only lend to those they knew were going to repay the loan. But reality has nothing to with this idyllic world of perfect information.

Until now, banks have been the most efficient system for matching supply and demand with the fewest possible information asymmetries. Their huge client bases enable them to reduce global risk of insolvency thanks to diversification, while allowing them to take advantage of the huge economies of scale underlying the activities needed to reduce information asymmetries. But corporations like Facebook and Google could become even more efficient than the banks in these activities, as they have larger client bases and have developed very precise routines for discovering the tastes and needs of their users, while also holding information about their incomes.[21] As these corporations are big data experts they may well have the most information in the world on the largest number of people. A priori there is little to suggest they could not act as intermediaries as efficiently as the traditional banks. Think about Facebook: 1400 million users voluntarily provide highly personal information about the things they like and what they do not like, the places they go to on holiday, the houses they live in, the jobs they do, the things they buy, and so on. Would it be possible to develop an algorithm to predict who is going to need a loan and how much he will need and the most suitable maturity date to minimize the risk of default? Gathering such information, which is very useful in deciding who to lend to and who not, is much more costly for a traditional bank. That is why the CEO at Spanish banking giant BBVA has declared that future competition will come from this kind of firms rather than other banks.

---

[21]In 2015, Facebook had 1.4 billion users, Google was estimated to have 300 million active users in Google+ (although the official figure is 2.5 billion) and Apple's iTunes had more than 800 million accounts, most with a credit card attached.

For this reason banks and MNO are ideal allies. The former for motives of defense, to be able to better protect their business as the economy goes digital and the latter because it enables them to enter value-added services that will improve ARPU by providing new sources of revenues and increasing user loyalty. Indeed, mobile money, as the set of financial operations that can be done on a smartphone is called, is a promising growth area in which carriers are major players. In three-quarters of the nigh-on 240 mobile money services in the world, there is a mobile network operator present (Almazan and Vonthron 2014). This is a field the MNO had already tried entering in the year 2000, but then the circumstances were not right for creating a viable ecosystem (Ernst and Young 2009). Before it could happen, substantial improvements were needed in smartphone and network technologies alike. Near Field Communication (NFC) chips, for instance, were an essential feature for smartphones. This technology, from the world of the IoT, was actually developed in 2003, but it was not until 2008 that some telephones were fitted with the chips. For the technology to be useful, and being a systemic innovation, point-of-sale terminals also had to be adapted. Stock trading, bill payment, inter-account fund transfer, P2P transfer are just some of the services lined up for major development in developed countries. Others are associated with mobile commerce like mobile payment using NFC technology.

Even in developing countries, where banking service accessibility levels are low, operators have found that mobile money has a place and is profitable. This is so, for instance, in bank-free rural zones with telephones, thanks to heavy price reductions, and with younger people, who are more likely to use new technologies. In such cases, this is an opportunity for carriers to lay the groundwork for other, greater value added services and to increase user loyalty. Vodafone, very active in Africa, obtained 14 % of its revenues in Tanzania through its subsidiary in the country via mobile money (Almazan and Volthron 2014).

In Spain in 2014, Telefonica allied with Banco Santander and La Caixa to enter the financial business with the Yaap Shopping platform. Like Santander, Telefonica is a major player in Europe and Latin America. It set up one of the first joint ventures with a bank. In less than a year more than 2000 establishments have joined the platform, which enables them to do high visibility, low-cost marketing, and run a digital shop window. Banks, carriers, commerce, and users all benefit from the platform.

6.5 Operators and E-Health

Healthcare expenditure is closely linked to age, income, and technology. Most of an individual's healthcare expenditure in his or her lifetime occurs after the age of 65. Some 60 % of people over 65 suffer some chronic illness. The populations of developed countries, particularly in Europe, are aging, and the process is set to continue in the near future. In 2020, 13.5 % of the world's population will be over 65 and in Spain, according to the UN, the figure will be 36 % in 2050. This factor brings the most pressure to bear on public expenditure. At present, Spain allocates 9.4 % of its GDP to healthcare, practically the average of OECD countries (9.3 % of GDP). At 16.9 % of its GDP, the US allocates most resources to healthcare. Most healthcare expenditure in developed countries is borne by the state (72 % on average in OECD countries and 73 % in Spain). Chronic illnesses like diabetes, high blood pressure, and asthma generate the greater part of the costs. Such illnesses need close monitoring and overload health systems by increasing the number of doctor's appointments.

Governments and healthcare experts alike are aware that the situation is not sustainable, and fully agree on the need to change the healthcare provider system model. It should focus more on prevention than cure, patient co-responsibility (self-care) should be encouraged and more intensive use made of information and communication technologies. Telefonica estimated that IT services meant major savings in healthcare costs (see Katz 2009). Low transmission costs support medical services. For instance, the digitalization of images versus film implies a 25 % saving, data digitalization improves nursing productivity by 50 %, telecare service save 80 % over hospital care, and telerehabilitation 9 %.

E-health is considered a solution to reducing the costs generated by aging and the accompanying chronic illness. The market is growing rapidly, and by 2017 is expected to reach $27 billion, with Europe the most important component (PWC 2013). E-health may well be an ecosystem benefitting patients, who can control their own health better, healthcare professionals, who have more information about their patients, thus facilitating diagnosis and treatment of their illnesses, and insurance companies, because e-health promotes healthier lifestyles and reduces service costs. 4G's high speed and data transmission capacity also facilitate personalized services, customized to the needs of users: the patient will receive individualized, more intense and immediate service because it helps to increase doctor–patient interactivity.[22]

All these services require the development of ecosystems involving specialist medical staff, public and private healthcare services, apps developers, handset and wearable manufacturers, and operators themselves. A host of healthcare-related apps (6000-plus) and wearables (e.g., smart pacifiers to detect if a child is running a high temperature or is ill) is already appearing, as are things like smart baby bottles, an oxygen saturation sensor, very useful for patients with sleep apnea and risk of heart attacks, chronic obstructive pulmonary disease, lung cancer, and asthma. Some wearables can warn patients a minute before they suffer a brain hemorrhage (e.g., Samsung), or help detect cardiac anomalies.

Operators have reacted fast and practically all the sector big names are in the business (e.g., TeliaSonera, Deutsche Telekom, AT&T, Verizon, Vodafone, British

---

[22]Orange has set up an e-health platform open to patients, doctors, insurance companies, etc., to provide a broad range of healthcare services: remote surgery, patient monitoring, online medical records, etc. More than 50 % of doctors in France use Orange systems to send patient treatment forms and the platform connects 12 million-plus French patients to their insurance companies (Friedrich et al. 2013).

Telecom, Telefonica, Orange, etc.) because the accompanying services consume large quantities of data that need a carrier to be involved. For instance, radiology images require a great quantity of data, and high-capacity and high-speed networks are best for sharing them. A high quality digital image of a chest X-ray uses 40–50 Mb (Maglogianis et al. 2006). They are, of course, highly sensitive data, as much so as financial data, meaning that reputation is a major factor in choosing the network. In this segment, MNO have a clear advantage over MVNO. Finally, the value added these services give to users is clear, so they obviously help reduce churn and increase ARPU.

6.6 Carriers and the Internet of Things

All reports agree on the upcoming gigantic growth in connected devices and machine to machine (M2M). Although figures differ, all are remarkably high. By 2020 the world is expected to house somewhere between 26 billion (Ericsson 2015) and 50 billion connected devices (Evans 2011).

Perhaps the most outstanding feature of the IoT is the objects' capacity to communicate using standard protocols, i.e., the IoT arose as the union of M2M with Internet. GSMA Intelligence (2015) reckons that of the 10 billion mobile connections in 2020, 974 will be M2M connections. In 2014 there were 150 million, which means accumulated growth of 36.6 %. Although this is a tiny fraction of the total, it is one of the biggest expected growth areas, so diversification helps operators revitalize global growth, compensating for the saturated mobile business, whose penetration rate has, as we have seen, reached saturation levels. Spectacular growth is forecasted for the IoT in a wide range of industries thanks to the combination of the currently low penetration rate with high potential profits for firms connecting the objects they manufacture. The IoT' disruptive potential in some sectors is very high, because it changes the rules of the game and alters business models, enhancing production efficiency enormously and value added for clients. By 2025 best-in-class organizations are expected to be making widespread use of IoT technologies in their products and operations will be up to 10 % more profitable.

As they have explicitly acknowledged in their annual reports, carriers' strategic objective is to enter the IoT. The logic behind this movement is obvious. First, there are many more objects that connect than people in the world, and most of them are not connected as yet. Second, growth in object connectivity should grow exponentially given the profits expected for firms that do so, the enormous value added they can generate and the almost unlimited positive externalities they will produce (e.g., reductions in greenhouse gas emissions, fewer accidents, better health, less inactivity, fewer hours lost in traffic jams or looking for a parking space, increases in productivity, etc.). And third, because such objects are connected to networks, and although there are many types of networks, operator-managed networks and, in particular, 4G are required for many IoT businesses (wherever large quantities of data are transmitted and low latency is needed, as in health or in-car safety). Telephone carriers are masters of the spectrum and also have the routines needed to manage networks efficiently. This is one of the fields where they are investing most, to maintain margins in the face of price reductions in voice and data.

Besides connections, communication between objects and between objects and people (P2M) requires maintenance and information transmissions between the two or between the device and the Internet. Any object, in principle, could be part of the IoT, which means the variety of types of devices that can be connected is practically limitless, which gives rise to a sweeping heterogeneity in information requirements and therefore in the shape of business models. For this reason, some IoT fields are more suitable for carriers than others. Diversity is enormous throughout four dimensions: (i) the movement of connected things; (ii) the amount of information connected things need to transmit; (iii) the scope of such information; and (iv) the speed at which they need to transmit it at for them to be useful. Some things are fixed, like electricity or gas meters, while others frequently move at high speed, like automobiles; some things send the information over short distances, and connect via Bluetooth, like payments with NFC technology, while others may need long distances, when, for instance, a container is monitored. Some things require high information transmission speeds, as in the case of illness detector devices or anti-accident devices; some do not generate much in the way of data transmission, like fire sensors, which might be inactive for years before having to transmit information on a fire, and when they do they use very few bytes. Others generate major information traffic, as in the case of those linked to a connected car. There are also major differences as regards network density requirements: devices used for road traffic, health, retail commerce, and finances require a very dense network, with heightened capacity. One of the major features of an automobile is, obviously, its mobility and users would find it very helpful for it to be connected at all times, which requires an extensive network.

Operators possess and manage large communication networks, putting them in an advantageous position to be top-level players in this development area, although given the huge variety of possible combinations, mobile carriers in principle will be more interested, and will enjoy more advantages in IoT businesses that require major network capillarity, abundant information traffic and high data transmission speeds. Healthcare, connected cars, financial services (means of payment), and home automation are areas in which operators are, for preference, taking up positions.

At present, only 3 % of connected devices use 4G networks because the majority do not need high speeds or large quantities of data transmission and 3G networks cover their requirements. Price reductions affecting LTE modems would prompt the appearance of new applications with very low latency requirements (Ericsson 2015), which is why 4G is expected to accelerate the development of the IoT.

Although the leading operators have entered the connected car in force, not all find themselves in the same position to appropriate the revenues generated. Rivalry between companies transfers from service to platform or ecosystem. Companies compete to put together platforms that users will find attractive. Not all carriers can do this, with only global operators having the clout required. This contributes still further to concentration in the sector through mergers and acquisitions or collaboration agreements. AT&T, the leading carrier in the connected car, with a 50 % market share in the USA has put together a connected car platform in which carmakers (Audi, BMW, General Motors, Ford Motor Co., Nissan, Subaru, Tesla, and Volvo) are involved as partners. Telefonica is also looking to build an ecosystem for the connected car with KPN. NTT Docomo, Rogers Communication, Singtel, Telstra, Vinpel Com and Etisalat, besides striking up agreements with carmakers (Tesla), insurance companies (Generalli), and so on. And Orange, Vodafone, and Telenor are part of the Automotive and Web Platform Business Group, set up by W3C to create a connected car web standard that does not yet exist. Also taking part in this group, alongside OEM from the automotive industry like Mitsubishi, Jaguar, GM, Ford, and Volkswagen, are Tier1 manufacturers, chip makers and browser and mobile service developers. Verizon's strategy deserves special consideration, as it appears to focus more on building an ecosystem by itself that will enable it to move along the value chain. Verizon has specialized in data transmission, one of the niches associated with the connected car that could generate most services and data. It has done so by external growth, by acquiring Hugues Telematics, a specialist in manufacturing voice and data connectivity systems for automobiles. Key to the success of this strategy is whether the existence of a technology standard is a decisive factor, or not, to be able to compete. If it were, then AT&T would have an advantage over Verizon, thanks to its bigger market share and because it has struck agreements with eight carmakers to use their connectivity technology, and this in part because it was able to develop a universal, remotely programmable SIM card for cars compatible with every network the world over. Verizon has also developed its own, but needs to install specific localization parameters for the country it assembles the car in. This could be a major factor in an industry where cars are designed in one country, are assembled in another and sold in yet another.

As an area for diversification, the connected car is very attractive for operators, and not just because of the prospects for growth (according to Gartner, there could be 250 million connected cars in the world by 2020). It also gives rise to a huge variety of services for the user, which the carrier can use to provide value added and thus increases client loyalty and ARPU. Connected cars generate a huge quantity of demand in data that have to circulate around operator networks and which therefore are susceptible to being converted into carrier revenues. The huge amount of data needs to be analyzed and stored safely, and thus cloud computing comes into the picture as a complementary development area to the connected car. So 4G is an opportunity for carriers to improve profitability and growth through related diversification.

6.7 Operators and Cloud Computing

Cloud computing is a major growth area. In 2008 this sector earned $46 billion, a figure that had risen to more than $150 billion in 2014, a cumulative annual growth rate of 21.8 %, and for 2016 it will top the $200 billion mark, according to reports issued by STL Partners (2012).

Growth here is also promising for operators; indeed, the big names are already investing heavily to position themselves in the sector. Cloud computing is divided into three major categories: IaaS (Infrastructure as a Service), SaaS (Software as a Service), and PaaS (Platform as a Service). Consultants Gartner (2013) reckoned that IaaS would have a CAGR of 41.3 % in the 2011–2016 period, PaaS, 27.7 % and SaaS 19.5 %, and would grow most in Pacific Asian countries and Latin America (31.8 and 25.2 %, respectively), although this particular market is biggest in North America and Europe.

With IaaS, the service provider makes servers, storage systems, routers, and other infrastructure features available to the client, together with applications. The provider is responsible for maintenance, backup, and security. Clients pay according to the use they make of the resources available to them in terms of time or amount of space used. This type of service enables client firms to avoid both investments in hardware and software and the associated technological risks. With this kind of system firms increase productivity and economic profitability, because the resources are highly scalable and they adapt to client demand, although they also hike the costs of change. Amazon Web Services is one of the main IaaS providers.

With Platform as a Service (PaaS), the provider supplies the infrastructure clients (developers of apps executed in the cloud) need to be able to concentrate exclusively on writing software and ensuring it is optimized to consume the fewest resources. These services greatly reduce complexity in the deployment and maintenance of applications. Some of the more important carriers are Microsoft (Windows Azure) and Google (Google App Engine), although Cloud Foundry has recently enjoyed major growth. The latter is an open-code platform also being developed by corporations like IBM (Bluemix), HP (Helion), and some thirty other firms in mid-2014.

Software as a Service (SaaS) is a software distribution model where programs and data are stored in the cloud, in the servers of an ICT company, and are accessed via Internet. For firms it is a way of simplifying access to data and software because they can do so from any part of the world and from any device, as they are not physically installed in a client's computer; access to them is gained through a browser. The provider is responsible for development, maintenance, updating, and backup copies, which means the client company saves on costs and investments in IT platforms and is left free to concentrate on developing the key competencies of its business. The downside is that the client will barely control the applications it handles. Examples of SaaS include Google Docs, Webmail (Gmail), online CRM, and Dropbox. All development, maintenance, updating, and backup copies are the provider's responsibility.

European carriers like Deutsche Telecom, France Telecom, TeliaSonera, Vodafone, and Telefonica offer SaaS services with Office 365, Libre Office, Box Web Meeting, Telefonica Mail, etc. For these operators it is a major source of differentiation that enables them to keep competitive factors of the price variable at arm's length, as confidence is one of the most important keys for success, and in this aspect reputation and brand are basic resources the big carriers possess. Digital service providers have to provide their client bases with securing data redundancy, support and backup. The progress of digitalization means companies are transferring an increasing number of services to the cloud, which in turn means the network has greater data transport capacity and faster data upload and download speeds, although they need to know they have access to their data at all times and that they are protected. In short, operators have to offer high-speed transmission and have a quality network. A cloud service slower than a local network would add scarce value to users (TeliaSonera 2014). Offering, for instance, videoconference services, firewalls, IDP, SPAM filters, remote M2M management services, and the like is enough for an operator to differentiate itself from others and avoid being seen as a dumb pipe.

After analyzing the 10 principal Telcos' track record in the cloud, STL Partners (2012) found that what most differentiates their behavior is the emphasis they place on their investments in being providers of infrastructure as a service (IaaS) or on being providers of software as a service (SaaS), and the decision as to whether to build or acquire cloud capabilities. They found that carriers grouped into one of three clusters depending on their behavior regarding two variables. The first, the farmers, is formed by Orange, AT&T, BT, Singtel, and China Telecom. They are large fixed telephony operators that preferred developing rather than buying Infrastructure as a Service. The second group, which STL calls the innovators, includes DTAG, China Mobile, and Telefonica. These corporations placed more emphasis on Software as a Service and tended to pursue more diversified technology strategies. Finally, the "gatherers," meaning Verizon Business and Vodafone, basically tend to pursue acquisition strategies, Verizon concentrating on the American IaaS market and Vodafone investing heavily in SaaS.

Services in the cloud are evolving rapidly and there is much uncertainty as to where they are headed. Depending on the road taken, they may continue to be hugely attractive to operators, or not. One probable evolutionary path is for services in the cloud to be dominated by a few corporations, including Amazon (EC2), IBM (Blue Cloud), Google (App Engine), and Microsoft, providing very generic services in public clouds, which take on gigantic proportions, but barely differentiated and which therefore would not be right for corporations or, for instance, the financial or governmental sectors, where the sovereignty of the data, their security, latency, and geographical distribution are crucial factors. Operators with LTE-A networks would then have the right conditions to satisfy the needs of such sectors left unsatisfied by public clouds. They could offer the financial sector very low latency services that meet the security criteria in data transmission and storage set out by PCI-DSS (Payment Card Industry Data Security Standard) (STL Partners 2012). Data security is one of the barriers that have made the public sector reluctant to use cloud computing services. Operators are favorably placed to provide G-Cloud services guaranteeing that data will be safe, secure and within national territory, sovereignty being a primordial variable for such organizations. But it is essential first to have deployed a fiber or LTE-A network. O2 and Iomart provide services in the cloud to the UK's public sector because they have data center networks connected by fiber within the country.[23]

7 Conclusions

Until now, in amongst the huge amounts of value added proliferating in the Internet ecosystem, the Mobile Network Operators (MNO) appear not to have found the path to appropriating a substantial part of it, despite heavy investment. Growth in their revenues has stagnated in recent years, largely because their principal markets are saturated and because of competition from virtual operators and Over the Top (OTT) firms also offering voice and data services using the networks deployed by the MNO. This situation is the result of epigenetic forces moving too fast for them and to which they have not been able to adapt at the required speed. Coevolution with the ecosystem is required if they are not to become marginal players.

4G is the latest foot-to-the-floor acceleration in the changes transforming the ecosystem, one that may leave the MNO still further behind unless they manage to adapt. Internet regulation, particularly the network neutrality principle, seems to go against them and for that reason they are pressuring for a change in the regulations. In Europe the situation is especially unfavorable for MNO because markets have not settled and jelled, the situation remains heavily oligopolistic in most countries, where just three operators have roughly 70 % of the market while in Europe as a whole more than 50 survive. In the face of such a fragmented and complex market, European MNO are at a disadvantage regarding their American and Asian counterparts.

MNO can appropriate revenues in the new environment shaped by 4G technology only by repositioning themselves in other parts of the value chain, but the sheer speed of the changes has given them no time to modify their routines, so they have chosen to react through mergers and acquisitions and collaboration agreements. 4G technology has speeded up mergers and acquisitions between operators and has improved consolidation in the sector. It has also triggered an enormous increase in the number of cooperation agreements with other players in the Internet ecosystem. In this sense at least, carriers have changed some of their old ways.

The Internet of Things (especially everything to do with the connected car, financial services and healthcare), cloud computing and the contents industry for

---

[23]See http://www.iomart.com/g-cloud-services/.

the ability to pursue service packaging strategies, are the most usual MNO strategic movements to counter this new epigenetic state of affairs ushered in by 4G. It is still too soon to tell if they are succeeding, as in virtually all these fields the fight is on for the right position, but as yet it's hard to appreciate consolidated positions in any of them. In some, like the Internet of Things, standards have yet to be developed.

References

Acquisti, A. (2010) The economics of personal data and the economics of privacy. Working paper. Carnegie Mellon University: http://cusp.nyu.edu/wp-content/uploads/2013/09/C03-acquisti-chapter.pdf. Access 5 Oct 2015.

Acquisti, A., Leslie, K. J., & Loewenstein, G. (2013). What is privacy worth? *The Journal of Legal Studies, 42*(2), 249–274.

Almazan, M., & Volthron, N. (2014). Mobile money profitability: A digital ecosystem to drive healthy margins. GSMA. http://www.gsma.com/mobilefordevelopment/wp-content/uploads/2014/11/2014_Mobile-money-profitability-Summary-Introduction.pdf. Accessed 5 Oct 2015.

Alonso, J., Sainz, C.C., De Lis, S.F., Tuesta, D. (2014). Una aproximación a la economía de los datos y su regulación. Observatorio Economía Digital. https://www.bbvaresearch.com/publicaciones/una-aproximacion-a-la-economia-de-los-datos-y-su-regulacion/. Accessed 15 Apr 2015.

Arlandis, A., & Ciriani, S. (2010). How firms interact and perform in the ICT ecosystem? *Communications and Strategies, 79*, 121–141.

Baird, Z. (2002). Governing the internet: engaging government, business and nonprofits. *Foreign Affairs, 81*(6), 15–20.

Ballestero, F. (2013). Las políticas públicas en el sector de las telecomunicaciones. *Papeles de Economía Española, 136*, 72–87.

Bijlsma, M., Straathof, B., Zwart, G. (2014). Choosing privacy: how to improve the market for personal data. CPB Netherland Bureau for Economic Policy Analysis. http://www.cpb.nl/en/publication/choosing-privacy-how-improve-the-market-for-personal-data. Access 5 Oct 2015.

Bock, W., Evans, P., Forth, P., Lind, F., Mark, D., Mei-Pochtler, A., Nill, C., Plaschke, F. (2014). *The 2014 TMT value creators report: productivity & growth*. The Boston Consulting Group.

Bock, W., Field, D., Zwillenberg, P., Rogers, K. (2015). *The growth of the global mobile internet economy: The connected world*. The Boston Consulting Group.

Broadband Commission (2013). The state of broadband 2013: Universalizing broadband. http://www.broadbandcommission.org. Accessed 5 Oct 2015.

Califano, B. (2013). Políticas de internet: la neutralidad de la red y los desafíos para su regulación. *Revista Eptic, 15*(3), 19–37.

Calzada, J., & Estruch, A. (2011). Telefonía móvil en España: regulación y resultados. *Cuadernos Económicos de ICE, 81*, 39–70.

Calzada, J., & Estruch, A. (2013). Comunicaciones móviles en la Unión Europea: tecnología, políticas y mercado. *Papeles de Economía Española, 136*, 29–49.

Carpenter, G. S., & Nakamoto, K. (1989). Consumer preference formation and pioneering advantage. *Journal of Marketing Research, 26*, 285–298.

CNMC (2014a). Análisis geográfico de los servicios de banda ancha y despliegue de NGA en España. http://www.cnmc.es. Accessed 5 Oct 2015.

CNMC (2014b). Informe económico de las telecomunicaciones y del sector audiovisual 2014. http://www.cnmc.es. Accessed 5 Oct 2015.

COM. (2005). *i2010: Una sociedad de la información europea para el crecimiento y el empleo.* Brussels: European Commission.

COM. (2010). *Una agenda digital para Europa.* Brussels: European Commission.

Commission, European. (2014). *Broadband markets.* European Commission: Digital agenda scoreboard. Brussels.

Curwen, P., & Whalley, J. (2011). Mobile telecommunications gives birth to a fourth generation: an analysis of technological, licencing and strategic implicatons. *Info, 13*(4), 42–60.

Ericsson (2015) Ericsson mobility report: on the pulse of the networked society. http://www.ericsson.com/res/docs/2015/ericsson-mobility-report-june-2015.pdf. Accessed 5 Oct 2015.

Ernst & Young (2009). Mobile money: An overview for global telecommunications operators. http://www.ey.com/Publication/vwLUAssets/Mobile_Money./$FILE/Ernst%20&%20Young%20-%20Mobile%20Money%20-%2015.10.09%20%28single%20view%29.pdf. Accessed 5 Oct 2015.

Evans, D. (2011). The internet of things: How the next evolution of the internet is changing everything. CISCO. https://www.cisco.com/web/about/ac79/docs/innov/IoT_IBSG_0411FINAL.pdf. Accessed 5 Oct 2015.

Federal Communications Commission. (2005). DA 05-543. http://hraunfoss.fcc.gov/edocs_public/attachmatch/DA-05-543A2.pdf. Accessed 5 Oct 2015.

Fernández, P. E. (2014). Neutralidad de la red: tensiones para pensar la regulación de Internet. *Question, 1*(42), 69–84.

Friedrich, R., Bartlett, C., Groene, F., & Mialeret, N. (2013). Enabling the OTT revolution: How telecoms operators can stake their claim. PWC. http://www.strategyand.pwc.com/media/file/Strategyand_Enabling-the-OTT-Revolution.pdf. Accessed 5 Oct 2015.

Fundación Orange (2014). *Informe anual sobre el desarrollo de la sociedad de la información en España.* Madrid.

Ganuza, J. J., & Viecens, M. F. (2013). Exclusive content and the next-generation networks. *Information Economics and Policy, 25*(3), 154–170.

Ganuza, J. J., & Viecens, M. F. (2014). Las subastas de espectro radioeléctrico en España. *Economía Industrial, 393*, 15–23.

Gerhardt, W., Kumar, N., Lombardo, A., Loucks, J., & Buckalew, L. (2013). Next-generation knowledge workers: accelerating the disruption in business mobility. *White Paper.* CISCO. http://www.cisco.com/web/about/ac79/docs/sp/Business-Mobility.pdf. Access 5 Oct 2015.

Gómez-Uranga, M., Miguel, J. C., & Zabala-Iturriagagoitia, J. M. (2014). Epigenetic economic dynamics: The evolution of big internet business ecosystems, evidence for patents. *Technovation, 34*(3), 177–189.

Grzybowski, L., & Liang, J. (2014). *Estimating demand for quadruple-play tariffs: The impact on consumer surplus.* Presented in 25th European Regional Conference of the International Telecommunications Society, Brussels (Belgium) (pp 22–25). June 2014.

GSMA Intelligence (2014a). GSMA Mobile Economy Europe 2014. http://europe.gsmamobileeconomy.com/GSMA_ME_Europe_2014_read.pdf. Accessed 5 Oct 2015.

GSMA Intelligence (2014b). Understanding 5G: Perspectives on future technological advancements in mobile. https://gsmaintelligence.com/research/?file=141208-5g.pdf&download. Access 5 Oct 2015.

GSMA Intelligence (2015). The mobile economy. http://www.gsmamobileeconomy.com/GSMA_Global_Mobile_Economy_Report_2015.pdf. Accessed 5 Oct 2015.

Han, Y. J. (2015). Analysis of essential patent portfolios via bibliometric mapping: An illustration of leading firms in the 4G era. *Technology Analysis & Strategic Management, 27*(7), 809–839.

Huertas, J. C. (2013). Fundamentos económicos de la regulación de las telecomunicaciones desde la perspectiva actual. *Papeles de Economía Española, 136*, 52–71.

iRunaway. (2012). Patent and landscape analysis of 4G-LTE technology. http://www.wipo.int/export/sites/www/patentscope/en/programs/patent_landscapes/documents/patent_landscape_4glte.pdf. Accessed 5 Oct 2015.

Katz, R. (2009). *El papel de las TIC en el desarrollo. Propuesta de América Latina a los retos económicos actuales*. Colección Fundación Telefónica, Cuaderno *19*, Madrid (Spain).

Lieberman, M. B., & Montgomery, D. B. (1988). First-mover advantages. *Strategic Management Journal, 9*, 41–58.

Little, A.D. (2015). Time to better leverage LTE: benchmarks, dynamics and insights into mobile data services. http://www.adlittle.it/uploads/tx_extthoughtleadership/ADL_TimeToBetterLev erageLTE.pdf. Accessed 15 Apr 2015.

Llobet, G. (2014). La licencia de patentes en estándares tecnológicos. *Economía Industrial, 293*, 51–58.

Maglogianis, I.G., Karpouzis, K., Wallace, M. (2006). *Image and signal processing for networked e-health applications*. Morgan & Claypool.

Ministerio de Industria, Turismo y Comercio (2010). *Estrategia 2011-2115: Plan avanza 2*. Gobierno de España.

Ministerio de Industria, Energía y Turismo (2013). *Agenda digital para España*. Gobierno de España.

Network Strategies (2013) LTE vs ARPU: data takes over. http://www.strategies.nzl.com/wpaper s/2013014.htm. Accessed 25 Apr 2015.

Page, M., Molina, M., & Jones, G. (2013). *The mobile economy 2013*. London: A.T. Kearney.

Pérez, J., & Badía, E. (2012). *El debate sobre la privacidad y seguridad en la red: regulación y mercados*. Barcelona: Fundación Telefónica-Ariel.

Pérez, J., & Olmos, A. (2008). Gobernanza de Internet. *Patronato de Fundación Telefónica, 80*, 58–62.

PWC. (2013). Diez temas candentes de la sanidad española para 2013: Para que la crisis económica no se transforme en una crisis de salud pública. https://www.pwc.es/es/publicacio nes/sector-publico/assets/diez-temas-candentes-sanidad-2013.pdf. Accessed 5 Oct 2015.

Rumelt, R. P. (1987). Theory, strategy and entrepreneurship. In D. J. Teece (Ed.), *The competitive challenge* (pp. 137–158). Cambridge, MA: Ballinger.

Schmalensee, R. (1982). Product differentiation advantages of pioneering brands. *American Economic Review, 72*, 349–365.

STL Partners (2012). Cloud 2.0: Telco Strategies in the Cloud. http://www.telco2research.com/ articles/SR_Cloud-2-telco-strategies-2012_Summary. Accessed 15 Apr 2015.

TeliaSonera (2014). The Telco & The Cloud. https://www.teliasonera.com/documents/Public%20 policy%20documents/WhitePaperOnClouedServices.pdf. Accessed 26 April 2015.

Thomas, D. (2015, February 2). Three chief Dyson questions consumer appetite for quad play. Financial Times. http://www.ft.com/intl/cms/s/0/3aceab78-aa0b-11e4-8f91-00144feab7de.html#axzz3d9iXsub9. Accessed 5 Oct 2015.

Van Valen, L. (1973). A new evolutionary law. *Evolutionary Theory, 1*, 1–30.

Wu, T. (2003). Network neutrality, broadband discrimination. *Journal of Telecommunications and High Technology Law, 1*(2), 141–179.

Scope and Limitations of the Epigenetic Analogy: An Application to the Digital World

Jon Barrutia, Miguel Gómez-Uranga and Jon Mikel Zabala-Iturriagagoitia

1 The Digital World

The development of digital ecosystems has undergone growth and an unusual transformation since its origins. This has created new business models, new ways to exchange information and new types of socio-economic relationships, all of which have been based on a core of continuous technological development. It is also important to note that the ecosystem is formed by several dimensions. One is related to communication infrastructures. Another is the network where the information travels. All the devices and/or resources that ensure connectivity and an accessible operating interface function between the two should also be considered. This third factor can be understood as an extension of one of the two previously mentioned dimensions. Besides, a new dimension needs to be included, namely, that which embraces socio-economics, which are global and interdependent, in scales that have never before been seen, going beyond political and geographic barriers, must also be noted. All that is taking shape within fast response times,

J. Barrutia (✉)
Department of Management and Business Economics, University of the Basque Country, UPV/EHU, Bilbao, Spain
e-mail: jon.barrutia@ehu.es

M. Gómez-Uranga
Department of Applied Economics I, University of the Basque Country, UPV/EHU, Bilbao, Spain
e-mail: miguel.gomez@ehu.es

J.M. Zabala-Iturriagagoitia
Deusto Business School, University of Deusto, Donostia-San Sebastian, Spain
e-mail: jmzabala@deusto.es

© Springer International Publishing Switzerland 2016
M. Gómez-Uranga et al. (eds.), *Dynamics of Big Internet Industry Groups and Future Trends*, DOI 10.1007/978-3-319-31147-0_8

which generate a new tempus for socio-economic actions (i.e. those carried out by companies, countries, organisations) and poses extraordinary innovation demands.

Technological advances coalescing as massive connectivity have enabled the Internet to become an authentically global phenomenon. However, the incessant increase in traffic calls for mobile bandwidth extension which, in turn, depends on the radio electric spectrum, which is a public good. Mobile connectivity (also fostered by the arrival of devices that facilitate it), together with new connections to machine to machine type networks, transport management systems, security systems or Smart grid devices point to an increase in user demand of mobile wideband (see Chapter "Future Paths of Evolution in the Digital Ecosystem" in this book by Pérez Martínez and Serrano Calle, and Chapter "4G Technology: The Role of Telecom Carriers" by Araujo and Urizar). Both the technological limitations and allocation of this public good already indicate a gap between demand and supply possibilities. This gap may create asymmetric geopolitical and socio-economic areas, as per the extent of network expansion and, as a result, may cause digital divides with the consequent inequalities they involve.

Technological and economic feasibility is needed due to dependence on infrastructures. This calls for efficient design and use of the radio electric network and its complementary options, or if applicable, its replacement. As quality differences between fixed and mobile networks diminish, the tension between the two is being resolved in favour of the latter.

It also necessary to note the incessant speed of product and services innovation as well as the constant and abrupt transformations that the various actors in the ecosystems and the ecosystems themselves are undergoing. Intense change and rapid innovation are constants in the digital (i.e. Internet) ecosystem. Partly as a consequence of the latter, new issues, not only institutional, regulatory and normative, but also related to security and privacy arise. They lead to the need to restrict companies' freedom via public regulation to safeguard citizens' freedom and open access to the network. All of these aspects are breaking down traditional business models and creating new models and value chains. The latter is now based on the digital "mode of production" with new parties and activities, content owners, online services, connectivity, etc. In these highly dynamic ecosystems, countries (i.e. environments) and firms interact in an operative manner. It is no coincidence that the USA and Apple and South Korea and Samsung are the global leaders in the digital world. We should not overlook the relationship between Apple and Samsung, which may also be partially explained by the epigenetic dynamics approach, as discussed in Chapter "Epigenetic Economics Dynamics in the Internet Ecosystem" in this book by Zabala-Iturriagagoitia et al. (see also Gómez-Uranga et al. 2014).

This alignment of countries and companies causes a certain amount of tension between agents concerning adequate network neutrality and related conflicts. Public regulation is the generic scope of resolution for these issues (see Chapter "Future Paths of Evolution in the Digital Ecosystem" in this book). From the perspective of globalisation, the digital ecosystem provides clear possibilities for an effective delocalisation at every possible level: operative, financial, commercial and even legal. Their scope goes beyond national States and laws, as they also modify the framework of global relations. Along these same lines of change, it has also transformed the information failures of traditional markets, shifting from information asymmetries to knowledge asymmetries. Accessibility conditions concerning capital markets or intellectual property protection can neither be overlooked. These aspects play a key role in these ecosystems and initially benefit big business groups.

Strangely enough, user empowerment is increasing in the same ecosystem, what has an important impact on the entire group of system actors. Therefore, gatekeepers generate the main initiatives and control in the configured ecosystems. In other words, they are the agents which determine access and participation in the system. It is the big established business groups (i.e. in this book we have referred to these as the GAFA) which compete to play the role of main gatekeepers and benefit from the corresponding privileges, thus causing problems related to lack of competition or collusion. Therefore, regulations are not only important per se, but their implementation and control have also become essential. Nevertheless, the difficulty involved is also extreme due to the trans-border and globalised nature of the digital ecosystem. These aspects question the freedom and security of countries, companies and citizens, leading to tension between the possibilities of greater freedom and violation of it in that intervention is allowed to an extent that had never before been possible. In short, the diversity of agents and operators, globalisation, regulation and security ultimately underscore the importance of Internet governance.

It is therefore necessary to understand the multi-stakeholder approach and note that an approach based on multilateral cooperation is not entirely valid on its own. For instance, it is not valid inasmuch as it limits the States' role merely to that of any other agent. They are placed on equal footing with the rest of the agents, which makes regulatory responsibility less effective. Nevertheless, there is a common demand for the Internet to be a global 'place' managed in the public interest.

We are ultimately facing a new world, like the discovery of America, where the amount and scope of transactions, as well as their economic value, social potential, the regulatory challenges and the new business models they called for and the global innovation they lead to are totally new to us because of their disruptive nature. The Internet constitutes a digital ecosystem where the pace of change is staggering, which makes the internal dynamic itself disruptive, demanding enormous adaptive capacities at all levels. Furthermore, it is important to understand that the type of knowledge being generated and the way in which it is applied to different realities causes it to change radically. In other words, it invades everything, ranking from traditional business models and industries to long lasting institutions, also causing disruptive changes in all of them. This speed of internal change is then transferred to the exterior (i.e. the environment), creating a rapid and double route of transformation in the economy. Dysfunctions, malfunctions or obstacles (to innovation) which cause different impacts arise when there is an imbalance between the two speeds, what in the context of the EED approach we have referred to as the consequences of epigenetic dynamics (see Chapter "Introducing an Epigenetic Approach for the Study of Internet Industry Groups" in this book by Gómez-Uranga et al.).

When the internal speed is much faster than the external, processes of acceleration may occur which lead to the elimination of certain agents from the ecosystem.

When the external speed matches the internal, traditional industries break down and new activities arise from old ones. These may have very different market valuations, which trigger relevant dysfunctions in the job market, etc. At the present time, health, financing, telecommunications or the higher education are areas that may undergo radical transformations in short as a result of the epigenetic moves of the large firms in the Internet ecosystem.

Nor can we overlook the fact that the digital ecosystem is formed by multiple subsystems, which count with different types and which are in continuous interaction. New phenomena, which need also be analysed, also emerge from these frictions and interrelationships, thus enabling us to understand the evolution of the ecosystem as a whole. More specifically, we might regard the GAFA (see Chapter "Epigenetic Economics Dynamics in the Internet Ecosystem" in this book by Zabala-Iturriagagoitia et al., and Chapter "GAFAnomy (Google, Amazon, Facebook and Apple): The Big Four and the b-Ecosystem" by Miguel and Casado) as two-sided markets; namely, companies that occupy a significant position on both markets and among citizens, with a strong innovation push, a global nature in their activities, all of which systematically influence the everyday activities of citizens, firms and public organisations of all ranges and types. Furthermore, they are the leaders in a complex ecosystem formed by reticular relationships which include different, yet also complementary, industries, activities and agents.

All of these form an ecosystem whose key characteristic is dynamic coexistence of organisations (i.e. the GAFA and their corresponding business ecosystems) engaged in hardware, software and content in varying proportions and in a systemic manner. Put differently, the GAFA function in a global and interdependent way so that they can provide consumers with higher value by offering the sum of the parts as a whole and in context (see Chapter "GAFAnomy (Google, Amazon, Facebook and Apple): The Big Four and the b-Ecosystem").

The conceptual discussion is useful and of interest, due to the implications it carries out. However, although each of the GAFA may be considered a business ecosystem (see Chapter "Epigenetic Economics Dynamics in the Internet Ecosystem"), the fact remains that the entire new digital communication world is a large ecosystem where agents set up as subsystems, which are directly interrelated. Thus, it is not only necessary to study the systemic dynamics of each of the GAFA but those of all of them as a whole, as they also go beyond the global level and overflow into other productive sectors and the entire economy. In this regard, Chapter "Epigenetic Economics Dynamics in the Internet Ecosystem" pointed out some of the consequences (i.e. malfunctions or dysfunctions) that the dynamics observed in the Internet ecosystem in general and on the analysed GAFA in particular have over a set of systemic elements such as fiscal policy, the treatment of intellectual property, competition, as well as social, cultural, or the rights of privacy of individuals.

Proof of this is that the GAFA mutually identify each other as competitors in the 10-K Forms they need to report to the United States Securities and Exchange Commission, and behave in this manner in the market game. They compete against each other for clients but also for sources of funding, or form alliances with other stakeholders to manufacture their devices, they provide incentives for developers to use their platforms and APIs, etc. Interrelation is thus extremely complex and interdependent, often of a type never seen before and with a high potential to overflow and devour resources, generating a new profile or version of the crowding-out effect (Aschauer 1989) in this case to other productive sectors.

Being a leader in this framework implies high innovation levels, what in turn requires highly dynamic and innovative capacities. To achieve this, and due to the fact that the endogenous rhythms of innovation processes of the GAFA do not happen to be flexible enough or sufficient, these groups are often forced to support their innovation processes and strategies from acquisitions, both from other companies as intellectual property rights of third parties. It should not be overlooked that the developers, a priori outside agents to the GAFA, also constitute one of the core areas on which the latter pivot their innovation strategies and decisions (see Chapter "The Digital Ecosystem: An "Inherit" Disruption for Developers?" in this book by Vega et al.). Finally, we should also mention the integration of new talent and competent human resources incorporated by these large business groups.

The competitive dynamics of the GAFA have some particular characteristics. Each works comes from being specialist in an industry (original routines and competences, which have often been regarded along the book as the original DNA of the business groups) where the others barely have any options to perform and grow. Thus, and to a certain extent, there is monopolistic competition. However, in the system as a whole, these firms hinder each other's mutual growth. Continuing with the epigenetic metaphor, the system is like a forest where each tree competes to have more space to grow. In this case, we might even say that the forest is formed by trees of different species, due to their different DNAs.

In this forest, the GAFA act as platforms where intermediation power becomes the strength to determine the "meeting point" between content providers and customers. As previously discussed, they are two-sided platform markets and as such, provide us with understanding about markets which are affected by global networks. This means an increase in the number of users on one market attracts those from another market. That is to say, the number of users in one group increases as the number in the other group does, creating a great incentive to form the highest concentration of activities (see Chapter "Epigenetic Economics Dynamics in the Internet Ecosystem"). On the other hand, they are complex platforms which, in turn, are connecting and intertwining, creating new ones, undergoing thus modular growth that strengthen the ecosystem rather than challenge it. Limitations derive from the urgent need to differentiate and the capacity and potential for diversification in this same process. The comparison with trees in the forest remains valid.

Competitive tendencies stimulate the closed (self-sufficient) nature of these systems and thus prevent the 'commodity system effect' (see Chapter "GAFAnomy (Google, Amazon, Facebook and Apple): The Big Four and the b-Ecosystem"), but their complexity makes them dependent on resources and therefore partially open. This is the case of developers, and particularly, the case of the Apple–Samsung partnership. The latter provides the former with strategic components and, as a result, has relevant information on innovation, quality and marketing strategies (see Gómez-Uranga et al. 2014). A cooperative and competitive relationship (i.e. often referred to as 'coopetition', see Brandenburger and Nalebuff 1996) is established between the two which also relates to the characteristics of epigenetic and systemic dynamics, as long as the two of them see themselves obliged to adopt certain decisions due to the pressure from the environment. In the same respect, we also find financial needs being strongly associated to the previous moves. The need for resources is so great that the tension between autonomy and openness is constant and dialectic. This complexity leads us to a context in which multiplier effect economies do not occur. Nor do economies of scale, synergies or cost economies in traditional terms although strong interrelationships and large dimensions do. It could be said these are 'symbiotic economies'.

A highly relevant amount of information is used and handled, knowing that their systematised digital mode management (i.e. the Big Data), leads to positions of knowledge and socio-economic actions which have been completely unknown to date. These have an enormous potential, not only for large public and private organisations, but also for small start-ups, developers and even individuals at large (see Chapter "The Digital Ecosystem: An "Inherit" Disruption for Developers?"). We therefore find ourselves, following our epigenetic approach, in a huge forest which is growing and expanding, in the competitive coexistence of different plant and animal species with their own biological legitimation processes, at least in their initial stages of development. If we would be able to show, with a certain temporal perspective, that this is the case in the Internet ecosystem as a whole, then it could be said that these dynamics would explain the different competitive natures of the epigenetic process in this macro-ecosystem. The macro-ecosystem has factors which are both material and immaterial, located and de-located, which end up creating a typical working space that penetrates the traditional economic space, thus giving it new forms and content.

In the field of telephony, as it has been discussed in Chapters "Future Paths of Evolution in the Digital Ecosystem" and "4G Technology: The Role of Telecom Carriers" in the book, the combination of telecommunications infrastructures and the computer world, which includes mobile network operators on the one hand, and over the top firms on the other, we find that there are truly asymmetric benefits due to positions of power and economic results, with the latter clearly winning out. It should not be overlooked that in addition to the millions invested by mobile network operators, they are subjected to much higher regulatory pressure than over the top firms.

From an epigenetic approach, it seems that the former have not evolved and adapted parallel to the latter. These asymmetries are accelerating with the arrival of 4G technology, which will force mobile network operators to make new moves (i.e. decisions) and adopt new dynamics if they are to avoid being excluded from the value chain, as they are obliged to gain a stronger foothold in the advent of the Internet of Things.

Ultimately, the market structure is far from being perfect competition. There are monopoly areas and many monopolistic competition areas, with the well-known entry barriers that are not only placed on economies of scale, but also reach knowledge agglomerations and the barriers to access resources. Disruptive innovation dynamics and their Schumpeterian relationship with socio-economic progress also arise in this context. Some of the points mentioned can also be illustrated by planned obsolescence as a strategy to achieve a certain level of innovative tension, which creates demands and excludes potential competitors. Nor can we overlook all the differentiation strategies that aim to create loyalty, which goes beyond the brand to reach operating systems and production logics. Oligopolistic logics are also found on these markets, with their corresponding strategic interdependence, tendencies to collusion and leader–follower dynamics.

The Internet being a business sector with a markedly horizontal activity, which affects many other sectors of activity and even infrastructures and institutions of the public interest, regulatory pressure is particularly high in some fields. In addition to other aspects, this issue creates exit barriers and institutional barriers that make the sector and its expansion even more complex. As per institutional barriers, it is important to note that companies create international environments in the digital world, making regulatory systems highly or even totally inefficient (see Chapter "Epigenetic Economics Dynamics in the Internet Ecosystem").

For all of the above, economic analysis usually falls somewhat short of offering explanations and predictability of the digital world. The epigenetic logic introduced in this book, and ultimately the ecology-biology analogy, may therefore be even more appropriate as a body of knowledge to offer explanations and predictability.

2 The Epigenetic Analogy and Digital Ecosystems

Digital and communication ecosystems form a new scenario on the world socio-economic panorama. It therefore does not only constitute an area in which the epigenetic analogy can be applied as an illustration of high-velocity markets, but it is also creating a new industrial revolution. In this respect, the environment-business ecosystem relationship, which our epigenetic approach is distinctive for (see Chapter "Introducing an Epigenetic Approach for the Study of Internet Industry Groups"), becomes highly dialectic in this context, and gradual iteratively mutual influencing phenomena occur. That is to say, the Internet ecosystem addresses the evolutionary demands posed by the environment, although in turn, it also creates new environments which call for new evolutionary demands and so on.

In this regard, it can be considered that there is a double creation of environments. On the one hand, there are intrinsic environments (i.e. those occurring within the business ecosystem formed by each firm), which are digital in nature and therefore closely related to the original production mode followed by the firms themselves. This scope would include the dynamics that involve mobile network operators (i.e. telecommunications companies) on the one hand, and digital firms selling telephony equipment, tablets, software firms, etc. on the other.

We also need to consider here the environments created as a result of the cross-cutting nature of the digital world's use or application in other economic sectors or industries. References are being made to the arrival and development of digital technologies in fields such as the manufacturing industry, health, education, mobility, construction, etc. The qualitative changes caused are usually so large in scale that they may even break down the traditional dynamics of these sectors. Cases in point are the Industry 4.0 and e-health. In this manner, the digital world exports its dynamics to other sectors and, makes them its own to a certain extent. It can thus be said that a third type of environment emerges and establishes the relationship between the two former ones. This third environment would actually form part of digital ecosystem, and therefore it should be included in analysis. In the book, this third environment has been partially included in the analysis of the consequences of epigenetic dynamics. However, as discussed in Chapter "Epigenetic Economics Dynamics in the Internet Ecosystem", the analysis of the consequences is still rather partial, as the consequences are only observed ex-post. Therefore, this remains a matter of further work.

One of the problems encountered in any analysis undertaken within the Internet ecosystem is the definition of the borders. Setting topological markers proves difficult due to the dynamic nature of the ecosystem concerned, its powerful permeability and penetration in other production sectors and its "eagerness" to radically transform these. It is a dynamic with limitless expansion, which means we could consider it as an open growing ecosystem. This is illustrated by the figure below, which follows, reconsiders and further develops the EED approach introduced in Chapter "Introducing an Epigenetic Approach for the Study of Internet Industry Groups".

This topological difficulty is transferred to the core issue of the strategic scope and industry configuration, in other words, the environment-organisation relationship. The environment may thus be considered to play the leading role as the driver of change, or this priority may be given to the substantial capacity of organisations to address or respond to changes in the environment. If the former is considered vital, environmental pressure is then understood to push selection and, as such, evolution (Hannan and Freeman 1977). This approach also implies that survival is based on successfully fighting to obtain the resources available in the environment (McKelvey and Aldrich 1983). In fact, very often the dynamics of the big Internet business groups are regarded as "the winner take it all", due to their enormous economies of scale, scope and network effects.[1] The profile or characteristics of the population under study are highly relevant from this point of view (Salimath and Jones 2011). Thus, one of the key explanatory variables to be considered in relation to the performance of the firms is the size of that population (i.e. number of firms competing in it). Population density and population mass (i.e. weighing each organisation by its size) are used as representative concepts in this regard (Bataglia et al. 2009; Freeman and Anderies 2015).

---

[1]The software industry in general, and the Internet ecosystem in particular, can be regarded as sectors where the main share of the total cost of the firms operating in them is a fixed cost. Accordingly, as the size of the companies becomes larger, the costs continue dwindling. In fact, since the marginal cost is almost zero, if there were perfect competition, the equilibrium price would also be zero.

When focusing on the population density and applying the density dependence model[2] (Hannan 1989), it can be said that legitimation and competition affect organisational dynamics from the perspective of firm survival. Legitimation is a sociological process which, in our case, confirms a certain type of organisation as the dominant one to face the demands involved in capturing resources in the environment, thus making it possible to take advantage of the opportunities offered. From this perspective, an active environment boosts the creation and proliferation of certain organisations over others. An increase in said organisations, in turn, enables them to capture the necessary resources more efficiently as synergistic effects are generated with the environment.

All of these issues fit squarely into the digital world, its arrival as a sector or industry in itself and as a cross-cutting application. Therefore, in other industries it is generating legitimation processes where organisational types share a minimum of key factors and, on the other hand, the number of organisations that emerge from this scope is also very high. In fact, it is difficult for firms in the digital world to escape from other competitive sectors. In the strictest sense of the digital ecosystem, namely, adopting a narrow view of the ecosystem in which only firms that belong to it are considered, legitimation processes like those described also occur.[3] The competition factor comes into play. In other words, legitimation enables a more efficient resource management. However, once the number of organisations and active players has grown to a certain number, and according to the evidence of the density dependence model, we reach an inflection point after which companies need to compete for the limited resources available (e.g., human, financial, technological). In other words, an U-inverted shape can be observed. We can also observe this aspect in the digital world, where dynamic legitimation processes are being accelerated over time.

The ability to determine the said number of active organisations could help to anticipate and/or understand some competitive dynamics. In this same respect, and simultaneously, organisations undergo concentration processes and make large acquisitions. This process is accelerated insofar as the size of the firm has a direct impact on its competitive strength (Winter 1990). Large acquisitions, either through processes of expansion, diversification, or mergers and acquisitions, create complex dynamics in which expulsion phenomena (i.e. the big fish eats the little fish) and cooperation phenomena occur, in which small organisations (i.e.

---

[2]This model and the following one are used in the field of Organizational Ecology to establish the competitive atmosphere and explain the mortality rate of organizations (Hannan and Freeman 1977; Hannan et al. 1995; Baum and Shipilov 2006). Organizational ecology is a valuable addition to the repertoire of theories that guide digital innovation policy when extended to community and ecosystem levels (Su 2011). Given the nature of the subject of this book, part of its findings and development may be applied to the EED approach introduced in it.

[3]As discussed in Chapter "Epigenetic Economics Dynamics in the Internet Ecosystem", an additional value of a device (e.g. tablet, Smartphone, etc.) increases the value of the previous device, insomuch as a general service is offered, at the same time, it works to increase the business ecosystem, which provides a substantial value in the product or service being offered. This is what Kelly (1998) labelled as the "fax effect".

start-ups and developers) gravitate, showing a large dependence, to the big ones. This would take us to the population mass scope. At this point, it would also be useful to know the impact of population mass on each sector and its direct relationship with legitimation and competition.[4]

From the organisational ecology perspective (Hannan et al. 1995; Becker 2007), and considering each organisation as a "test" to address the demands in the environment, it seems logical to think that the more tests, in other words, the more organisations there are, the higher the likelihood of successful realities when effectively addressing environmental pressure. To some extent, this is occurring in the digital world and its different "worlds", which would be those created as the digital world is introduced in other sectors. The capacity for success also depends on the extent of its organisational and productive logic. The more organisations and dynamism there are, the higher the probabilities of "finding" the most appropriate organisational types.[5]

An alternative point of view gives higher prevalence to the intrinsic capacity of organisations as competitive strength, rather than to pressure from the environment. This implies that organisations' functions, routines and resources (i.e. dynamic capabilities in the sense of Teece et al. 1997; Teece 2007, 2012) are what make them capable of continuing to properly nourish themselves and form combinations. Within the genetic code expressed in the original routines of the business groups, certain aspects that contribute to a proper evolution may arise, either from a phenotypic or a mutative perspective, with aptitudes for plasticity in rapidly evolving environments (i.e. high-velocity markets), and which allow an adequate relationship with relevant stakeholders (see Chapter "Introducing an Epigenetic Approach for the Study of Internet Industry Groups"). Finally, it cannot be overlooked that this world is party to a new industrial revolution and, as such, creates a new environment. As has been stated in this chapter (see Fig. 1), we are in front of a global and open ecosystem in a continuous expansion.

There are new analytical challenges for the biological analogy in general, upon which the field of evolutionary economics is based, and for the EED approach introduced in this book in particular, that require the construction of models with

---

[4]In relation to the density dependence model, a mass dependency model can also be found (Barnett and Amburgery 1990). From this point of view, the size or the organizational dimension is understood as the main variable explaining the competitive behavior, so that firm competitiveness is positively related to size. However, far from density and mass being alternative explanations, they may be compatible and can improve our understanding of the sector's behaviour. Along these lines, there are very interesting developments in Moyano and Nuñez (2002, 2004) applied to sectors different to the digital one, which make the two models compatible.

[5]As far as the present study is concerned, the most appropriate focus may be synthetic, in which the external approach (i.e. the environment induces change) can be made compatible with the internal one (i.e. internal capacity determines evolution). Given its complexity, formed by its own cross-cutting dynamics, the digital world creates fields to test the two approaches simultaneously. However, it is more difficult to track the endogenous or internal approach in comparison with the external one. The latter has some quantitative indicators such as density and mass that make it possible to run reasonably accurate analyses.

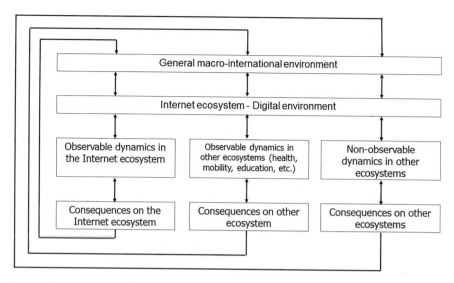

Fig. 1 Reframing the EED approach. *Source* Own elaboration

explanatory and predictability potential so that they can orient policy makers' and business management's decision-making. This last aspect is an even more sensitive subject and calls for further research, which we aim, as a community, to be able to accomplish in the following years to come.

References

Aschauer, D. A. (1989). Does public capital crowd out private capital? *Journal of Monetary Economics, 24*, 171–188.

Barnett, W. P., & Amburgey, T. L. (1990). Do larger organizations generate stronger competition? In J. V. Singh (Ed.), *Organizational evolution: New Directions* (pp. 78–103). Newbury Park: Sage.

Bataglia, W., Silva, D., & Meirelles, E. (2009). Population ecology and evolutionary economics. *Management Research: Journal of the Iberoamerican Academy of Management, 7*(2), 87–101.

Baum, J., & Shipilov, A. (2006). *Ecological approaches to organizations. The sage handbook of organization studies.* London: Sage Publications.

Becker, F. (2007). Organizational ecology and knowledge networks. *California Management Review, 49*(2), 42–61.

Brandenburger, A., & Nalebuff, B. (1996). *Co-opetition: A revolutionary mindset that combines competition and cooperation in the marketplace.* Boston: Harvard Business School Press.

Freeman, J., & Anderies, J. M. (2015). The socioecology of hunter–gatherer territory size. *Journal of Anthropological Archaeology, 39*, 110–123.

Gómez-Uranga, M., Miguel, J. C., & Zabala-Iturriagagoitia, J. M. (2014). Epigenetic economic dynamics: The evolution of big internet business ecosystems, evidence for patents. *Technovation, 34*(3), 177–189.

Hannan, M. T. (1989). Competitive and institucional processes in organizational ecology. In J. Berger, M. Zelditch, & B. Andersen (Eds.), *Sociological theories in progress: New formulations* (pp. 388–402). Newbury Park: Sage.

Hannan, M. T., & Freeman, J. (1977). The population ecology of organizations. *American Journal of Sociology, 82*(5), 929–964.

Hannan, M. T., Carroll, G. R., Dundon, E. A., & Torres, J. C. (1995). Organizational evolution in a multinational context: Entries of automobile manufacturers in Belgium, Britain, France, Germany and Italy. *American Sociological Review, 60*(4), 509–528.

Kelly, K. (1998). *New rules for the new economy: 10 radical strategies for a connected world.* New York: Viking Penguin.

Mckelvey, B., & Aldrich, H. (1983). Populations, natural selection, and applied organizational science. *Administrative Science Quaterly, 28*(1), 101–128.

Moyano, J., & Nuñez, M. (2002). Demografía organizativa y supervivencia: estado actual de la investigación. *Investigaciones Europeas de Dirección y Economía de la Empresa, 8*(3), 45–58.

Moyano, J., & Nuñez, M. (2004). El tamaño de la población como determinante de la probabilidad de desaparición organizativa. *Revista europea de dirección y economía de la empresa, 13*(1), 11–24.

Salimath, M. S., & Jones, R, I. I. I. (2011). Population ecology theory: Implications for sustainability. *Management Decision, 49*(6), 874–910.

Su, D. (2011). Review of ecology-based strategy change theories. *International Journal of Business and Management, 4*(11), 69–72.

Teece, D. J. (2007). Explicating dynamic capabilities: The nature and microfoundations of (sustainable) enterprise performance. *Strategic Management Journal, 28*, 1319–1350.

Teece, D. J. (2012). Dynamic capabilities: Routines versus entrepreneurial action. *Journal of Management Studies, 49*(8), 1395–1401.

Teece, D. J., Pisano, G., & Shuen, A. (1997). Dynamic capabilities and strategic management. *Strategic Management Journal, 18*(7), 509–533.

Winter, S. G. (1990). Survival, selection, and inheritance in evolutionary theories of organizations. In J. V. Singh (Ed.), *Organizational Evolution: New Directions* (pp. 269–297). Newbury Park: Sage.

Printed in the United States
By Bookmasters